Mental Toughness Secrets

Develop Daily Habits to Build Mental Toughness, Willpower, and Self-Confidence to Achieve Your Goals!

Written By C. Baker

© Copyright 2019 by C. Baker - All Rights Reserved

© Copyright 2019 by C. Baker - All Rights Reserved

This document is geared towards providing exact and reliable information concerning the topic and issue covered. The publication is sold with the idea that the publisher is not required to render accounting, officially permitted, or otherwise, qualified services. If advice is necessary, legal or professional, a practiced individual in the profession should be ordered.

From a Declaration of Principles which was accepted and approved equally by a Committee of the American Bar Association and a Committee of Publishers and Associations.

In no way is it legal to reproduce, duplicate, or transmit any part of this document in either electronic means or printed format. Recording of this publication is strictly prohibited and any storage of this document is not allowed unless with written permission from the publisher. All rights reserved.

The information provided herein is stated to be truthful and consistent, in that any liability, in terms of inattention or otherwise, by any usage or abuse of any policies, processes, or directions contained within is the solitary and utter responsibility of the recipient reader.

Under no circumstances will any legal responsibility or blame be held against the publisher for any reparation, damages, or monetary loss due to the information herein, either directly or indirectly. Respective authors own all copyrights not held by the publisher.

The information herein is offered for informational purposes solely and is universal as so. The presentation of the information is without a contract or any type of guarantee assurance.

The trademarks that are used are without any consent, and the publication of the trademark is without permission or backing by the trademark owner.

All trademarks and brands within this book are for clarifying purposes only and are owned by the owners themselves, not affiliated with this document.

Table Of Contents

Introduction

Chapter 1: Fundamentals of Successful Rituals

- Learn to discover your talent
- What is fear
- The scale of the foundations of successful rituals

Chapter 2: Rituals of Highly Successful Individuals

- Morning rituals of eleven highly successful individuals
- Evening rituals of five highly successful individuals
- Corporate rituals of three highly successful individuals

Chapter 3: Strengthening and non-power Rituals

- The power of your thoughts
- Nine common disempowering rituals
- Six enhancement rituals

Chapter 4: Building Successful Rituals

Chapter 5: Successful Six-Core Rituals

- Physical health
- Emotions
- Relations
- Career or business
- Finance
- Spirituality

Chapter 6: Morning Success Rituals

- Smile, smile
- Show gratitude
- Drink more water
- Positive statement
- Exercise
- Plan For Tomorrow

Chapter 7: Evening Success Rituals

- Enhance the evening questions
- Capturing your magic moments
- To celebrate

Conclusion First Part

Second Part

- Goals that never happe

Chapter 8: The most powerful skill you can learn: Setting goals The problem with your Current Goals

- What good goals look like
- How to formulate your goal

Chapter 9: The formula: How to Structure the Goals and Implement your Plan

- Step 1: display
- Step 2: Evaluate your situation honestly and completely
- Step 3: formulate a plan
- Step 4: Phrase your goals in a small step

Chapter 10: Let Go of Fear

- Setting the fea

Chapter 11: How to Achieve your Fitness Goals

- How to set and stick to realistic goals
- Adapt
- Have fun

- Play your strengths
- Take it eas

Chapter 12: How to Achieve your Career Goals

- Knowing what you want
- Creating a crazy-proof strategy
- The path of least resistance
- The quick Fail mode

Chapter 13: How to Achieve your Relationship Goals

- Make the point
- Creating a schedule for appointments
- Knowing what you want

Chapter 14: How to Achieve your Travel Goals

- Alternative travel strategies

Conclusion Second Part

Bonus Book!

- How to do more in a fraction of the time Wor

Chapter 15: Priorities

Chapter 16: Tips to Help you set Priorities

Chapter 17: Beat Procrastination

Chapter 18: Tips to Stay Focused

Chapter 19: Working Less Perform More

Chapter 20: Self-Confidence

Chapter 21: Use of Statements and Conclusion Finals

Introduction

You want to become successful. You can feel that there is more to life than what you are presently enduring. It has crossed your mind more than once that maybe you are doing things the wrong way. You need a new approach to ensure you give life your best shot at becoming one of the individuals that history can never forget. You can make your indelible mark on the line of history as a successful person by trying some Success Rituals.

You are right about needing an approach to attain success. Success Rituals shares insight into steps that successful individuals conduct daily and some that you might want to implement in your life.

The natural order of life is that things go through stages. Take for instance the process that it takes to create a child. In the womb, there are a series of steps, and after birth, there is another. Nevertheless, each child develops at a different pace. Some kids walk or talk before what is considered the "expected time" for them to do so. Success Rituals are processes that individuals endure to become prosperous. Achievements during the stages are different for everyone. However, there are similarities in each stage of the Success Rituals, just as how there are similarities in the stages to produce a child, but the outcome differs.

Success Rituals have certain fundamentals that every successful individual had to do, and some of which they had to learn how to overcome. Likewise, you must decide without any form of doubt that you too will have to work hard and overcome whatever obstacles you will encounter on your journey.

You might be the one that will develop before the expected time frame for one to become a success. Do not delay for another minute. It is time to discover empowering Success Rituals, which will assist you to conquer your fears and progress on your destined journey of prosperity.

The pace and intensity of our lives, both at work and at home, leave several of us feeling like a person riding a frantically galloping horse. Our day-to-day incessant busyness — too much to do and not enough time; the pressure to produce and check off items on our to-do list by each day's end — seems to decide the direction and quality of our existence for us.

However, if we approach our days in another way, we can consciously change this out of control scheme. It just requires the courage to do less. It may seem simple, but doing less can be very difficult. Too often we mistakenly believe that doing less makes us lazy and results in a lack of productivity. Instead of doing less, it helps us enjoy what we do. We learn to do less of what is foreign and engage in less self-injurious behavior, so we create a rich life that we feel really good about.

We learn to do less of what is extraneous and engage in fewer self-defeating behaviors, so we create a rich life that we truly feel great about.

Just doing less for its own sake can be easy, startling, and transformative. Imagine having a real and unhurried conversation in the middle of an unforgiving workday with somebody you care about. Imagine completing one discrete task at a time and feeling calm and happy about it. In this book, you will see a new approach.

The approach is equally useful for our personal life and our work life. The two hemispheres of our work and personal lives constantly reflect on and affect one another, each changing and/or reinforcing the other.

Every life has awesome meaning, but the fog of constant activity and plain bad habits can often obscure the meaning of our own.

Acknowledge and change these, and we can again enjoy the ways we contribute to the workplace, enjoy the sweetness of our lives, and share openly and generously with the ones we love. Less busyness leads to appreciating the sacredness of life. Doing less leads to more love, more effectiveness and internal calmness, and a greater ability to accomplish more of what matters most to us.

Introduction Bonus Book!

How many incomplete goals do you currently have on your agenda? If you're anything like the vast majority of us, then chances are that you have hundreds of projects that you started and never completed, countless goals that you told your friends but never saw through and all kinds of dreams that seem to be getting less and less likely to come to fruition.

And it's for this reason, that you may find people roll their eyes when you tell them your 'next big project'. When you start a new training program to lose weight and everyone – including you – knows that you're likely to have lost interest by month two. Or when you talk about the app you intend to make, the website, or the business project. Or when you talk about that dream trip to Japan…

This is the way of things for many of us. We work incredibly hard at things we don't feel passionate about just to put food on the table but when it comes to fulfilling our dreams, we are remarkably ineffective.

It's time to change all that and to start making those goals happen. But how can you turn it all around?

How We're Going to Fix Your Goal Setting and Help You to Start Living the Life of Your Dreams

Accomplishing goals is about strategy, it is about making a cognitive shift to change the way you're thinking and it's about being smart about how you approach each goal. It's also about knowing how to choose your goals and even how to phrase them.

This book is going to show you how to make those changes then. You'll learn how to choose and write goals effectively, how to write effective action plans and how to make sure you stick with your goals and never give up.

But this book is going to be a little different than most goal-setting tomes, too. After we've given you the broad tools you need to start setting and accomplishing your goals, we're then going to take a look at how you can begin to put them into practice.

Because while a goal can be pretty much anything, for many of us they are going to fall into one of a few different categories. Most of us have goals for our relationships, goals for our fitness, goals for our careers and goals for travel. We're going to provide not only the abstract strategies you need to start making effective goals then but also the step-to-step processes that will let you apply these strategies in each of these areas. By the end of this book, you'll be adept at setting and accomplishing any goal. And at the same time, you'll have powerful strategies for improving your relationships, your fitness, your career and more.

Ready to change your life?

Chapter 1: Success Rituals Fundamentals

Do not be scared of the word rituals. Ritual is just a synonym of the word habit. We all know that our thoughts and actions over some time will determine how we progress in life. Positive rituals yield positive results, and negative rituals will produce undesirable results. We have many rags to riches story, and you might be the story of a rag awaiting the right rituals to change your situation to riches. You will never know until you examine your life and make the necessary modifications. It is time to change your perception about rituals.

- **Learn To Discover Your Talents**

You will find individuals with similar Success Rituals, but they all produce different results. This is because everyone has different opportunities and how we react to situations is different. Notably, persons will have similar talent, but each talent comes with a unique ability. If you do not discover the uniqueness that comes with your talent, you might never truly stand out from the others that are in the exact field as you are. If you cannot produce extraordinary results, then your advancement will be either mediocre or below average.

Mediocre and below average results will never give you the title of a successful person.

You will only discover the uniqueness that comes with your talent after you have started using your talent. Let us take a look at some Success Rituals Fundamentals that you must ensure you develop. Build your rock-solid Success Rituals starting with those listed on the ladder below:

The first step on the Success Rituals Fundamentals is to believe in YOU. If you do not believe in yourself, you will never really discover your true potentials. Apart from finding what hidden treasures of skills you have buried within, is knowing, how and when you are at your best. Some people will tell you that they think better having a hot cup of coffee. Some after they take a shower or go for a jog, then ideas will start to overflow in their heads. Think back to the time you were able to unravel your best idea. What were you doing at the time that happened and where were you when that happened?

If possible, go back to the exact location and repeat the same action of what you were doing, and try to discover more great ideas. You will need to purchase a notebook. You must always have

something close by to write down your thoughts. I am suggesting that you get a notebook because ideas will pop in your head at various hours of the day or night. For example, you might find yourself suddenly awakened by an idea that popped into your head, and you want to write it down immediately.

Moreover, trust me, this does happen in the middle of the night when you start developing yourself. Electronic devices take some time to get booted up, and you might lose your thought. However, if you have a notebook you just quickly fetch your notebook and make jottings, then in the morning, you could type and save your concepts in whatever storage software you use. No matter how brilliant you are, you will not be adept to retain all the concepts in your head. The thought that you will forget, if you do not jot it down, might be the very idea which is your ticket to get on the success train.

Developing the healthy habit of jotting down your ideas also allows you to analyze better your plans, and to be able to group or tweak them. That is progress, and you will feel good at what you have accomplished and be more motivated to put them into action.

You are elated now because you have chosen what you believe is the best idea and you are ready to put it into action. It will be at this moment that fear will opt to cripple your mind. Creating doubts about your idea and your ability to implement them successfully. If you do not believe in yourself, and if you have not entirely convinced yourself that this is your time. That you were born to make an indelible mark on the lines of history and you are feeling your greatness churning in the depths of your guts. That you cannot control it anymore and you need to find your best life. Then, I am afraid that you will only get stuck at jotting down great ideas and not knowing where those ideas would propel you on your predestined path of prosperity.

- **What is Fear?**

Fear is the mastermind behind the death of many goals. We all know someone who is an "If I had." The "if I had" of life have many ideas and are always the first ones to tell you how to get things done even though they have never tried anything worth mentioning in their lives. Yes, they have all annoyed you and me at some point in time in life.

You might have even been an "if I had," but the fact that you are reading this book means there is hope for you. It is not too late for you to decide to change your situation in life.

The only way to dispel your fear to take the necessary steps to set realistic goals and timelines for your concepts. The moment you decide to continue with your ideas you will feel the power fear had over your mind slowly losing its grip. Fear will entirely lose its grip on your mind when your concept becomes an action. So go ahead, even if you have to move with wobbling legs and shaking fingers, take your leap of faith.

- **The scale of the foundations of successful rituals**

Goal setting and time management are equally important on the Success Rituals fundamentals ladder. If you do not set targets with timelines, then you will have nothing to work towards and no achievements to look forward to celebrating. Setting goals is a medium that you can utilize to determine how close you are to achieving your plans. You can assess and adjust where necessary. It also assists in developing your skills set and discovering other talents that you were not aware that you possess.

Hard work and determination is what will transform your goals into reality. If you only work hard some of the time, then, you should anticipate the partial result. A mind that is determined is the fuel for hard work, and hard work keeps you determine. That is why both of them have to be used together consistently. It is impossible to achieve by having only one of the two. Think of determination and hard work that they are Siamese twins. They should always be joined, and it must be difficult to identify one from the other. Muhammad Ali has a quote that has an element of humor to it but it also clearly depicts what you have to do if you want to achieve your goals.

"The best way to make your dreams come true is to wake up." – Muhammad Ali

You have to wake up, be determined, and work hard.

Making improvements to your skills set and plans, are another set of Siamese twins. If you improve on your skill sets, then you will discover other things that you can accomplish. You can gain success in other areas other than what you had made plans for initially.

If you improve your strategies, then the possibility is that you will discover new abilities that you had no idea you had.

The discovery of new abilities can be by attempting to work on the improved areas of your plan yourself, or you can employ someone who completes the task in your revised proposal. If you hire an individual to perform the work, this is also a learning experience for you.

You must always be aware of what is happening in whatever business transaction that you had initiated.

For me, the final Success Rituals Fundamentals is never fear failure. You will inevitably fail at some stage of your plan. Failure varies for different individuals. You might experience a massive failure, and have to start your journey entirely from the first step. Your failure might be minor and with a few adjustments, things can be back on the progress level. If you fear failure, you will not be able to recover from misfortunes. When you face failure, accept it, analyze what is the possible cause or causes, and start strategizing immediately to correct, as well as, prevent it from reoccurring, your fears will begin to diminish.

Success Rituals are crucial because we all need them to become successful. If we try to skip a step, then, we can be sure of a definite fall. The fall will be in the form of failing to achieve our goals. Take things to step by step while you discover yourself and find your sole purpose for being awarded the gift of life. Learn to enjoy your roller coaster ride that life is guaranteed to take you on. No matter how badly you fail, you can always raise again if you work hard and if you are determined enough.

Chapter 2: Rituals Of Highly Successful Individuals

Are you wondering if Success Rituals are real? Well, this is what this section will clarify for you. I will be sharing with you rituals of high success and how it will help you get ahead of the regular folks in life. Take a look at the morning, evening, and business rituals of some highly successful individuals below:

- **Morning Rituals Of Eleven Highly Successful Individuals**

Mark Zuckerberg (Co-founder and CEO of Facebook) – I am wondering have you noticed that the man who has his name enlisted on almost all the billionaire list that you can grace your eyes on seems to only have gray T-shirts. These gray T-shirts are not a uniform for his company. The wearing of only gray T-shirts by Mark Zuckerberg is a deliberate act. This is his way of saving time in the mornings. Even after he is sometimes up all night discussing with an employee, you can still find Mark Zuckerberg up by the clock strikes six in the morning. Without out having to worry what clothing to wear for the day Mark Zuckerberg grabs his outfit which is usually his gray T-shirt and then he is off to work at his office early in the morning.

Padmasree Warrior (CEO NextEV, U.S.) – A true warrior of time and rising early in the mornings to complete most of her task is Padmasree Warrior. While the night shadows cover the land, and with the twinkling of the stars which probably reminds her that she is among the most successful women; at 4:30 in the mornings Padmasree Warrior is up and working. Padmasree Warrior starts the day by going through her emails for approximately an hour. Then, she ensures she is kept informed about current affairs by reading the newspaper. After her reading of the papers, then, it is time to ensure she remains fit and healthy, so she exercises. After her daily morning rituals, she is in her office by 8:30 in the mornings, ready to take on the challenges of a new workday.

Tim Cook (CEO, Apple) - Tim Cook not only enjoys his race of being up before the sun, but he is very proud of the fact that he is the first one to be at his company in the mornings and the last one to leave during the evenings. You can check for an email from Tim Cook as early as 3:45 in the morning; because that is when he gets up, and has become known for sending company emails at that time in the mornings. Tim Cook is one who ensures that he maintains his health; therefore, you can find him in the gym by five in the mornings.

Jack Dorsey (Co-founder, Twitter) – Tweet, the tweet might be some of the twittering that Jack Dorsey hears at 5:30 in the morning when he is taking his six miles jog. Jack Dorsey also takes some time to meditate before he leaves for his run in the mornings.

Jack Ma (Founder, Alibaba Group) – As precious and swift as the wind is the commodity time. We cannot preserve time, and once past, we can either give thanks that we had used our time productively or live with the regrets of what we never used our time to accomplish. Knowing how precious time and his family are Jack Ma is up by the latest seven in the mornings. Jack Ma uses half an hour to complete some task, and then he ensures he spends some quality time with his family.

Kara Goldin (Founder & CEO, Hint Water) – It is time to take a hint. Most highly successful individuals are hitting the work button before the day is dawn, and Kara Goldin is among those who are doing so. Kara Golding day begins at 5:30 in the mornings. She peruses through her work calendar ensuring that she has no exigent meetings, and then she responds

to emails. By 7:15, Kara Golding starts making her business calls, but not before she saturates her taste buds with a double latte, and she goes hiking with her husband.

David Cush (CEO, Virgin America) – Crunch time is at 4:15 in the mornings for David Cush. His fingers get busy to dial his associates' numbers who are on the East Coast, but not before he took some time to send his emails. Next, David Cush tunes his ears to Dallas Sports Radio, while his eyes are kept busy reading the newspaper. Then, he is off to the gym to ensure he keeps himself fit.

Dan Lee (Director, NextDesk) – It is time for a standing ovation because Dan Lee is up by the clock strikes 3:30 in the mornings. Dan Lee ensures that he is completely hydrated by drinking two liters of water, then, he also consumes two cups of coffee, and smooth things off with a smoothie. After Dan Lee hydrates his body, the next ninety minutes, he shares with his dog and reading. He reads for one hour and spends half an hour with his dog. Clearing his pours through perspiration, you can find Dan Lee in the gym from 5:15 to – 6:15 in the mornings. When the clock ticks its way to 7:15 in the mornings, Dan Lee is already in his office, prepared to tackle the challenges of a new workday.

Sallie Krawcheck (Co-founder, CEO Ellevest) – It is the romancing of the mind with the lights dimmed or sometimes she seeks warmth from the fireplace, along with a hot cup of coffee and Sallie Krawcheck is ready to start her day at four in the mornings. While the lights might be dimmed in her home at four in the mornings Sallie Krawcheck is shining bright with ideas; because that is when her creative flair comes to light best.

Indra Nooyi (CEO, PepsiCo) – Popping as early as four in the mornings, is, Indra Nooyi. Her first task is ensuring that her plans are organized for the new day. Indra Nooyi is usually buzzing in her office by seven in the mornings.

Richard Branson (Founder & Chairman, Virgin Group) – He is certainly no virgin when it comes on to rising early. Not even the comfort of his private island can prevent him from pulling his curtains at 5:45 in the mornings to watch the rising of the sun. With a fantastic view of the rising sun, which comes with the opportunity to enjoy his fortune for another day, Richard Branson maintains his health by exercising and having a healthy breakfast. Then, he is off to work to ensure that he keeps making his billions.

- **Evening Rituals of Five Highly Successful Individuals**

Bill Gates (CEO, Microsoft) – Bill Gates, dubbed the richest man in the world according to the Forbes Billionaires list. He reads for an hour almost every night before going to bed, no matter how late he gets home. One of the topics that have consistently sustained his reading habit is business-related issues, which Bill Gates uses to assess changes in the market (no surprise there right). He also read about politics and healthcare.

Joel Gascoigne (CEO, Buffer) – Walking for twenty minutes every evening. During his walk, Joel Gascoigne assesses his workday, analyses his greatest challenges, and then, he will slowly stop thinking about work when the shadow of tiredness takes over his body.

Arianna Huffington (Founder, Huffington Post) – At nights Arianna Huffington disconnects from the world of technology by turning off all her electronic devices. Then, she dissolves the stress from her workday by taking a hot bath. Arianna Huffington pulls the shutters on her day wearing her pajamas while she reads a physical book.

Kenneth Chenault (CEO, American Express)

– Kenneth Chenault gets a head start to his days by setting goals for three things he wants to achieve every night before he goes to bed.

Oprah Winfrey (Business Woman and Media Mogul)

– Oprah Winfrey ends her days just as how she starts them. She meditates twice per day, once in the mornings, and once in the evenings

- **Corporate rituals of three highly successful individuals**

Amancio Ortega made it to Forbes Billionaire list as the second wealthiest person in the world. Amancio Ortega has five business rituals, which have guided him to success.

Speed is very important

– Amancio Ortega took the retail industry by storm in 1975 when he founded his company named Zara. He used the tactic of ensuring that two times per week his store was restocked with the latest fashion. Another one of his strategies was getting new styles of clothing before his competitors, and the timely processing of the customer's order within forty-eight hours.

Amancio Ortega's plan worked; because he was able to satisfy his market better than his competitors, and have earned his organization the name of "Fast Fashion."

Obsess with your customers need

- One of Amancio Ortega's motto is that the clients are the driven force behind a business, and should always be the central focus when designing your business. Customers need must also be your focal point when you are deciding on the operational systems that you will implement at your business, style of clothing that you will be selling, and any other activity that requires you interacting with your customers.

Being customer-focused meant that Amancio Ortega conducts consistent market research. His detailed analysis included him observing fashion blogs and directly garnering information from customers to keep a brief of the current trends in clothing.

Being In Control Of Your Distribution Channels - Amancio Ortega capitalizes on the cost-effectiveness of China's clothing, but he also imports most of his products from other regions such as Morocco, Spain, and Portugal. Designing and sewing his products is one of medium Amancio Ortega swiftly meet the needs of his market by supplying them with the latest trends in fashion. He also utilizes a local network of sewing shops by having his designs cut, and treated in mills, then, they are sewed by the local sewing shops.

Being Committed To Your Roots

– His ears buzzing from the ideas his employees share with him, as he sits, and work alongside them is the only office that Amancio Ortega has ever had. Amancio Ortega is from humble beginnings. He is the son of a housemaid, and railway worker, but his desire for a better life led him to stop attending school at the age of fourteen to start earning. Knowing what it is like to have nothing, Amancio Ortega never got himself an office, but instead, he takes a hands-on approach by working with his employees. Age has not slowed him down; because even at the age of eighty Amancio Ortega still goes to his office on most days.

Continuous Innovations – Complacency has no place among those who want to progress, and for those who want to remain successful. To become complacent is the biggest mistake you can make. You must have either grow or die attitude, and if you want to be innovative, then, you cannot focus on the results.

Carlos Slim Helu has the tag of the fourth richest man in the world according to the Forbes Billionaire list. He has ten business rituals.

Making money in a downturn – Recessions cannot hinder Carlos Slim Helu from acquiring new businesses. His strategy is to take advantage of the companies that have been affected by the economic crisis, which has no significant financial problems. Some of the best companies are sold for half their values during a recession. Carlos Slim Helu was able to purchase the largest cement company world Cemex during the recession. Cemex currently has a net worth of approximately six billion.

Simplicity in organizational structures is best – Having a simple organizational structure with minimal hierarchical levels will allow

executives and lower line employees to be able to interact more frequently, sustain flexibility, and assist in quick decision making capability.

Remain focused on innovation, growth, training, and quality

– You must focus on innovation, growth, quality of products, training for employees and continually improve production processes. Analyze your organization based on global benchmarks, seek the most cost-effective means when possible to reduce expense, and increase productivity as well as your competitiveness.

You must live without fear and guilt –Carlos Slim Helu believes fear is the worse weakness that men can have. Fear first weakens you, then, it impedes your action and eventually leads to depression. He believes that guilt is a terrible burden in people's lives, which influences the way one thinks and acts. Guilt and fear create difficulty for your present-day, and it is a hindrance to your future. To overcome both fear and guilt one must have good sense, accept ourselves as we are, which means with whatever virtue, disappointments, and realities.

Making wise Invest in areas that customers find it difficult to avoid you – Carlos Slim Helu has made investments in a variety of industries. He has invested in the health sector, clothing industry, real estate, bakeries, telecommunication, academic institutions, and museums just to name a few. Carlos Slim Helu's many investments have allowed him to serve his customers daily through various mediums, making it impossible for them to conduct business, and not contributing to one of his establishments.

A good education will assist you to manage big business better – Carlos Slim Helu believes that good decision making facilitates business success. You will be able to make better business decisions if you are armed with the information to do so, this comes from increasing your knowledge through education.

***Try to be humble no matter your* status** – Carlos Slim Helu is aware that riches also come with a lot of responsibility; however, this has not prevented him from spending time with his family. He assigned two days out of each week for quality time with his family. One day for dinner with his sons, and another day for dinner with his twenty-three grandchildren. Carlos Slim Helu with all his billions, still drives himself to work, which is approximately an hour away from his home. He has also lived in the same home in the same neighborhood for over forty years.

Prepare yourself for big opportunities – Carlos Slim Helu believes one of the characteristics of becoming a successful entrepreneur is having the skill to identify great business opportunities and capitalize on your chances.

Committing to the game – Entrepreneurship and investing is like a game. If you want to succeed, then, it requires you to become committed to your entrepreneurial process and make decisions as if you are playing to win your entrepreneur and investing game to achieve success.

Comprehending your business to the core – Allegedly, Carlos Slim Helu controls approximately two hundred companies in different regions around the globe. He was able to accomplish this by ensuring that he comprehends his business to the core.

David Koch, the tenth billionaire on the Forbes Billionaires list. David Koch shared his five tips for you to start learning from those who are rich.

Below are David Koch's five tips:

Creating wealth through earnings and not saving –
Successful individuals like David Koch's continuous thoughts are how to earn big money and how to expand their potential profits. If you are not significantly increasing the sum of money that you have, then, there is a good chance that your savings will not make you a wealthy person.

Never be afraid to believe in your ability, and you should take smart risks – Many individuals refuse from taking risks because of the possibility exist that they might fail. However, those who are successful know that you have to take risks both financially and in your personal life if you want to earn significant rewards. Successful individuals also accept that failure is the price you pay for your ultimate learning experiences and you must develop the confidence to continue after you have failed.

Do not become emotionally attached to money –
Prosperous individuals take a rational approach when they are building their wealth. They do not allow any negative emotions like anxiety, greed or regret to daunt their financial decisions, which have led to better chances for them to become successful.

***Capitalize on all your opportunities*–** Successful individuals understand that every opportunity presents the possibility for them to achieve more, even if it is a partnership, project, or they are just negotiating to venture into a new business. They understand as well as appreciate the significance of networking. Successful individuals are always seeking new business ventures while capitalizing on their current assets to generate more income from different areas. Their positive attitude in every opportunity contributes to achievements in their business ventures and also to their wealth.

***Understanding that your time does not equal money*–** Dispel the belief that the time you spent working hard will be equivalent to your level of success. Undoubtedly, successful individuals do work hard, but they have worked even smarter. It is not necessarily how hard you work but how smart are you working. Smart working means you strategize your time wisely, how you utilize your assets for them to assist you to increase your earnings. You should aspire to become an expert in your field, no matter what field of business you are working.

If your greatest desire is to disengage yourself from the grip of poverty, then you should adopt the rituals of the wealthy, be fearless, and have absolute confidence in your abilities. Believe that you deserve to have all the best that life has to offer.

Chapter 3: Strengthening and Non-Power Rituals

It is never too late to find new rituals or to improve on those you are currently practicing. We all have rituals, but the question that you need to ask yourself is: "Are my rituals empowering me or are they disempowering me?"

How to Make a list of the daily rituals that you have been doing over the past ten years. Analyze the outcome of each of these rituals. From your analyst of the habits that you have developed over time, how did they assist you in progressing to your desired has your daily routine affected your relationship with your family and your friends?

If the answers to the above questions are ones, which clearly state that your habits have generated only a negative or mostly negative impact in your life, then it is evident that you have disempowering rituals. It is time to change those disempowering rituals to empowering rituals.

- **The Power Of Your Thoughts**

You have the power to change your thoughts, which will eventually influence your actions. Every one of your actions came about because of a thought that you had initiated. Whether that idea came about consciously or it was a subconscious thought. To create empowering rituals you will have to learn how to reprogram your mind. To accomplish the reprogramming of your thoughts you first have to find an empowering ritual that you will be used to replace the disempowering one.

Start your day with the right attitude, full of energy and refreshed with ideas. When your day gets off to a good start, you will be able to use your time more productively. We all have the same twenty-four hours in our day but how we structure our activities for the day can make a lot of difference.

Let us take a look at some disempowering rituals that you might be practicing, and were not aware that they are not empowering you.

- **Nine Common Disempowering Rituals:**

Getting out of bed long after the sun has comfortable settle itself in the sky and have provided great warmth for the Earth.

Lack of physical and mental exercise.

Unhealthy eating habits, which are leading to degenerative and other fatal health issues.

Your time is used for reading and watching the wrong kind of materials, which cannot assist you to achieve your ultimate goals.

The individuals that we interact with daily influence our thought process a lot more than we often realize. You will not find any successful people hanging out on the side of the street speaking to a group of individuals that only hang out on the street every day unless they are doing some kind of community outreach program. Gossiping on the telephone for hours would be considered unnecessary spending and valuable production time lost for a successful individual. If you keep spending most of your time with unproductive people, then, eventually you will become sterile. Make sure that you are not wearing the label of an unsuccessful person. Evaluate your acquaintances. If you associate yourself with people who are thriving for success or who are already successful, then, you will be motivated to change your lifestyle to be more in line with the successful people whom you are now associating yourself with.

Now, I am not saying that you should stop speaking to your friends if they are not productive. I am merely suggesting that you associate with them less and find more individuals that share similar success vision as you do. Successful people associate with other successful individuals.

Never making any plans for your day, you just go with the flow of any events. If you are not making plans to achieve, then you have quickly set yourself up for failure. So do not be surprised when you reap failure's results.

Poor time management. You might have made plans for your day, but you spent too much time conducting rituals that unproductive.

No form of academic improvement. While it is true that most of the world's wealthiest individuals might not have a master's degree, they continually educate themselves through daily reading of books or by watching various videos that will assist them in improving their skill sets.

The poor allocation of your finances is a disempowering ritual. Investing your money in things that cannot return a profit on your dollar will not let you become rich. It is nice to have a fancy car, beautiful clothes, and dine at the best restaurants, but these must only be a reward for the assets you have a mass through your hard work. Living like the rich and famous when your bank account along with your wallet is still living in poor man's land, is a perfect example of you allocating your funds poorly. All the finer things of life that you want to enjoy now will still be available when you have mastered your success rituals, and you are financially capable of living life as a wealthy and famous individual.

- **Six Empowering Rituals**

Exercise - your greatest wealth will always be your health. You should never sacrifice your health to achieve wealth. Your wealth will never truly restore your health, and you might not even live long enough to enjoy your success.

You must eat healthily and ensure you have adequate rest.

Get up early in the mornings. A great ritual that you must develop is to get out of bed early in the mornings. Creating a great head start to your day and life begins with you getting up early.

Make plans in your journal for the following day before you go to your beds at night and evaluate the activities of your day.

Find time to unwind and meditate. Spend some quality time with those you love.

Invest in yourself by increasing your knowledge about the field that you are in, and also learn how to spend your money wisely. Consistently, assess how your current asset can be allocated to allow you to earn more in different areas. Here is an excellent food for thought as you make plans to educate yourself:

"Formal education will make you a living; self-education will make you a fortune." Jim Rohn

What you do daily influences your life. If your rituals are disempowering, you will never yield positive or favorable results. On the contrary, if your rituals are empowering, then, you have a greater possibility of achieving some form of reward. Success starts in your mind, with the thoughts that you are forefront in your head because those thoughts are what will eventually empower your daily actions. Take a bit of advice from a woman who has dragged herself from the gutter of poverty to stand proudly among the wealthy, *Oprah Winfrey* stated:

"What we dwell on is who we become."

Chapter 4: Building Your Success Rituals

After you have identified the rituals that you currently have, your next step is to assess what rituals you need to practice. Then, it will be time to build your success rituals.

You cannot build your success rituals if you are not aware of what is that you need to be doing to ensure you become successful in your given field. If you want to become successful, then you must find out what rituals those who are successful in your given field have. Then, based on the information that you have garnered, you will build your success rituals.

I believe it can be comfortably said that the majority of successful individuals are always out of bed before the sunrises. With that said, the first ritual that you should build is ensuring that you are also out of bed before the sun rises. A guru of productivity *Benjamin Franklin* stated:

"Early to bed, and early to rise makes a man healthy, wealthy, and wise."

Let us examine the benefits of rising early in the mornings:

There is just something about the dawn of a new day before the sun makes its appearance in the sky that fills your inner vessel with a lot of hope. Not even the polluted air, which is often found in urban regions, can eradicate that morning whiff of hope, even if you just open a window in your home to take a swift, deep breath of it.

A second benefit of rising early in the mornings is that you do not have a lethargic feeling, which typically comes with rising after the sun has risen.

You can get a lot more done during a day by having an early start.

The peacefulness of the morning gives you a chance to filter your thoughts more accurately and make better plans.

So you have mastered the task of getting up early, that is great; however, you must use your time productively. The best part of becoming wealthy is to know that you had done so while maintaining or with no comprise to your health.

To maintain or not to compromise your health, you will have to watch what you consume and ensure that you do at least half an hour of physical exercise each day or for a minimum of three days per week. Most of the successful individuals have incorporated exercise into their daily routine; therefore, the next morning ritual that you should develop is exercise.

Exercise is important to get your body in shape, but it also helps with developing a healthier mind for you to function better. Exercising, as you know, is only a part of the process to keep healthy. You also have to eat healthy if you want the exercise to work efficiently.

Exercising and eating healthy is one way that you might cheat death to live a little longer so you can enjoy your wealth. So I am sure it is worth taking the shot at having a healthy lifestyle.

The aspiration to become the best at what you do is something that you must take into consideration when you are building your rituals. The great inventor and co-founder of Apple *Steve Jobs* said:

"*Innovation distinguishes between a leader and a follower.*"

You should consistently brainstorm for ideas that will improve the plans you already have, and what will uniquely set you

 apart from others. Whoever is the best in the market will also amass the most success. Allocate time to cover all areas of your life. You need to find the time to exercise, meditate, make your schedule, and work. You need time to have fun with your family and friends.

Start creating your opportunities for success by ensuring you have some highly successful individuals in your circle. You already know that success is something that you have to work to obtain. So you have to not only think like those who are best but start talking, dressing and acting like you are the best too. That is a sure way of setting up yourself to become the best in your given field.

Chapter 5: Successful Six-Core Rituals

There six core areas of success. Each area is important to create balance in your life. If you ignore one of the six core areas of success, then, you will not be able to function at your best. You will have a sense of lacking in your life and might even waste your time trying to fill the void with the wrong things.

The six core areas of success are listed below:

- **Physical Health** – The emphasis can never be too much on how important it is to ensure you do all that is possible to sustain your physical health. It is quite logical that one of two things will occur if you do not maintain your physical health. It is either that you will have to spend your fortune on medications and doctors or you will be snatched too early death due to some form of health issue. Exercise, eat healthily, get adequate sleep, and drink a lot of water can assist in keeping you healthy.

- **Emotions** – Your emotions affect your mind. If you are mentally unhealthy or unstable, then you cannot make objective decisions. A decision that is made when someone is emotionally unstable can wreak havoc on your life and has the potential to become very drastic, which we have all either heard, read or experienced personally.

All six core areas of success are correlated. Take for example, if something happens to trigger your emotions negatively – you are feeling sad, or you might be angry - you can always exercise to calm yourself down. It is not only your negative emotions that you need to learn to control because if you are too excited or happy, but you can also make the wrong decision. For example, if you are overzealous you might spend money on things which you do not need, and that money could have been invested in something that will increase your income. Life is about creating the right balance in everything you do.

- **Relationships** – Your relationships can affect your health and your emotions. A toxic relationship will leave you with feelings of despair and anger. Unhappiness and rage can lead to depression. Depression will affect your progress, either by you losing time to work or by you making decisions, which will ultimately fail.

The benefit of good relationships will create a heaven for you on earth. The world in your eyes is at peace because your heart is full of love, and your mind will have its cover made from thoughts of joy. You will function better. You will be more eager to get up in the mornings because you are grateful for another day to be with the ones you love.

Examine your relationships, and see how they are impacting your life. If being around someone makes you feel burdened, or you feel as if that person is pulling all your hope, and joy out of you, then that is a toxic connection. You need to disconnect from that person. Relationships that motivate you to become better or the ones that help your ideas sparkle are what you need in your life.

- **Career or Business** – Observing the attitude people display at their place of work can always tell who loves their job from those who do not. If you are not in the job or business which makes you feel that this is what you were born to do, then you might be in the wrong field. Your career or business must leave you feeling fulfill no matter the obstacles you face daily. The belief that the world could not exist without you doing that business or career finds its resting place in your mind and heart. When in the right field, it will be easier to keep focus, and because of the passion you have for your career or business, failure cannot convince you to quit.

- **Finances** – *"Money isn't the most important thing in life, but it's reasonably close to oxygen on the "gotta have it" scale."* – *Zig Ziglar*

Wealthy and happy is all of us heart's desire. However, if you want to be wealthy, it might take years of hard work to become rich, but it only takes a minute with a bad investment, which will result in you losing all your money. If you do not have money, it might make you unhappy because you cannot buy the things you need to support yourself or your family.

Not having money can also prevent you from investing in your career or business. However, you can have money and still not be happy, because there is unbalance in one of the other five core areas of success. Never spend money on things just because you want to impress others. That is not a ritual of successful individuals. The wealthiest people live humble lives.

For example, Bill Gates has topped various billionaire lists for years. He can afford his private jet. Bill Gates is known to comfortably sit in the economy class on airplanes when he is traveling.

Warren Buffet, with all his billions, is very much content in his home, which he bought for less than forty thousand dollars many years ago. He also still makes his billion-dollar transaction discussion on his flip phone, which he has not replaced for any of the high-tech cellular phones that are available today.

They give back a portion of their wealth to various charitable organizations. Sometimes it is a charity that they have started or one that is already in operation. Even after death, the richest individuals are sharing their fortune. A lot of affluent individuals are willing their fortune to charities. Melinda Gates and Bill Gates have a grant-making foundation where they give away billions to different charities around the world. They have also encouraged other billionaires to donate some of their fortunes to the less fortunate.

You must also adopt this principle of giving back to the less fortunate. Just as how you will start practicing the other success rituals, this is one habit you must also develop. Allocate whatever you can afford to charity now, and as your wealth increases, then, you increase the portion for charity as well.

- **Spirituality** – No matter what your religious preference may be, spirituality is an important aspect of spiritual success. Spirituality can be found all around you, from the time you spend in nature, to meditation, to your religious practices.

Spirituality can ground you and keep you close to both your personal feelings and help you work through anxiety and emotion which will enable you to make sound and important judgments. Choose a spiritual ritual for yourself and stick with it every day.

Chapter 6: Morning Success Rituals

A vast amount of information was shared about morning success rituals throughout this book, and it is evident that you have to start your day right, to create the perfect momentum for the rest of the day.

- **Smile, smile**

Your first exercise routine should be a smile. A smile is free. It is the best cosmetic surgery, and it will help to relax you. Try smiling right now, and see if you do not feel an instant peace in your heart.

- **Show Gratitude**

After you smile, then you should show gratitude to the world around you for the ability to experience a new day. Through meditation, you will hear your inner voice more distinctly. You will have the opportunity to search your soul, and discover your most profound desires. Sometimes, if you listen keenly, you will begin to understand exactly what your path should beDrink More Water

Drink some water at least a liter to hydrate your body. Water helps with circulation, improves your skin tone, assists with weight loss, and the purification of your body. Water is the best drink to have. You must always consult your physician about things that will affect how your body functions, before attempting to do them.

- **Positive Affirmation**

You need positive affirmation. Motivational audios or books are a great source of motivation. After you meditate, then, you can listen or read from something from your favorite motivator. Positive affirmation is a must among your daily routine, and you have to learn to be your greatest source of inspiration. This will help to block all the negative voices, which will tell you that you cannot achieve it.

Feeding your mind with positive food sets the tone for you to improve your knowledge. Take some time in the morning to read. Most of the knowledge you will acquire in life will come from your experiences and what you teach yourself.

- **Exercise**

Next, you engage in some vigorous exercise. Go for a long walk or a run. Do some stretches or yoga. Get physically active.

By the time you have completed the above activities, you will feel fully rejuvenated, and ready to face the new day. You can start with your most challenging task, and then work your way through the others. How you end your day can impact how your day starts. Therefore, aim to end your day on a good note.

Your body needs time to unwind from the wear and tear of the workday. Take some time to relax in the evenings. Designate time for your family. Never neglect the ones you love in your pursuit of wealth. It will be very lonely at the top of the ladder success if you do not have anyone to share your success with.

- **Plan For Tomorrow**

Make plans for the next day. Prioritize your task. Make a list of those that are critical to least important. Assess your day if you had accomplished your entire task, and if you had not, what prevented you from achieving that task. If it is possible, you should exercise in the evenings too.

Say goodbye to the world of work with whatever relaxation techniques you have and settle in for a healthy prosperous future.

Chapter 7: Evening Success Rituals

Now I will share with you the evening success rituals for going to bed every night feeling fulfilled with a big smile on your face! These rituals are the follow-up to your morning rituals to reflect and celebrate after your day's activities.

While morning rituals are highly encouraged to be incorporated into everyone's daily routine, the evening rituals are equally as important to get the most out of your day. After all, there is no point to start a race full of energy but no idea how to end it.

So, I don't want you to miss this crucial part to inspire your success and celebrate it. You may be consistently crushing your goals one after another, but unless you take the time to celebrate every victory in your life, then you're not getting the most out of your day and missing out on a lot of magic moments and sense of fulfillment that you should be experiencing at this moment.

Here are the Evening Success Rituals that I highly recommend incorporating into your evening routine:

- **Empowering Evening Questions**

At the beginning of the Evening Success Rituals, the main focus is to reflect your day. And the best way to reflect is to ask powerful questions, not just any other questions that beat yourself up. I called these powerful questions "Empowering Evening Questions", designed to help you reflect on your entire day and come up with constructive ideas to shape a better tomorrow.

So the first thing you should do is find a quiet place, where you won't be easily disturbed and distracted for a set amount of time to go deep in your daily reflection.

The questions you ask determine what you focus on. That's the reason why the type of questions you ask yourself is extremely important. Ask a lousy

the question and you will feel lousy; Ask a good question, and you will feel amazing!

For example, how you feel about 'Today' is ultimately based on the Good or Bad you focused on.

Let me ask you this... What were the things that happened today that you can feel good about?

I'd bet you can come up with a long list of things! It could be the fact that you're alive, or you went for a walk, or you had a wonderful time with your loved ones, or you had done a good deed, had an amazing dinner, crossed off ONE stuff on your to-do list… Whatever it is, you can always find something to be feel good about every single day.

Now let me ask you again… What were the things that happened today that you feel awful about?

Again, I'd bet you can list out a lot of things. It could be that you procrastinated on your tasks, ate some nasty junk food, failed to pick up that call, skipped the gym, said some awful things to people you care for, and so on… Ultimately, how'd you felt after answering this question? Of course, AWFUL!

The fact is, it is always Good and Bad in each day, and what you ask yourself decides what you'll focus on each night before hitting the sack. What you focus on determine how you feel about that particular day.

Since it is always Good and Bad, why not be proactive and decide that you only want to acknowledge the Good that happened that day? Since you can decide how you're going to feel at the end of the day, why not choose to feel good instead of bad?

Guess what happens if you constantly ask yourself Empowering Evening Questions? Every day will be an awesome day for you and you'll always feel like a rockstar! Now, think of the level of accomplishments, productivity, and fulfillment you'll experience every day… I'll let you know about mine - Phenomenal! There is no such thing as a bad day.

It all starts with your focus, and these Empowering Evening Questions will help you to do just that.

And here's a list of questions that I ask myself every evening:

What was fantastic about today? What did I learn today?

What am I grateful for today?

What was my biggest accomplishment for today? What would make today great?

Don't just write down all your answers, try to FEEL the emotions and energy coursing through your veins as you read out your answers out loud.

Do this with level 10 intensity - move, gesture and smile widely as you answer them. Allow yourself to feel proud, excited, happy, loved, appreciated, etc… Trust me, you'll be amazed by what this simple exercise can do to your physical and emotional well-being. The more emotional intensity you put into this exercise, the more juice you'll get from the Empowering Evening Questions exercise.

- **Capturing your magic moments**

A life worth living is a life worth recording. - Jim Rohn

After you've gone through the Empowering Evening Questions exercise, you'll be in an incredible state. But you don't want to stop right there. You should take this opportunity to seize these emotions and feelings by journaling all your successes, accomplishments and magic moments. Otherwise, they'd be forgotten.

So take out a journal and write down everything that happened on that day. List down everything that you had done and all the powerful moments that you want to remember later on.

Time passes by so fast and we easily forget to take in and appreciate the little things. At the end of your life, you won't be able to remember everything but only at certain moments. So I want to encourage you to take your time and capture all your successes and magic moments into your journal.

One day, when you're feeling down and see no hope in life, you have a journal to go back to and remind yourself of all the magic moments that you once had and realize that your life is more incredible than you think. By journaling your magic moments, you'll be aware of what you were doing with your time and celebrate your victory.

- **To celebrate**

Last but not least, CELEBRATE! Celebrate your day to your heart content, you deserve it! This is a powerful way to positively reinforce yourself.

Don't beat yourself up for things you didn't do, tasks you procrastinated, the food you shouldn't eat, etc… Because that's what most people do, which prompts them to feel awful and guilty about themselves. To make things worse, they repeat this process every day and they end up in a downward spiral that sets themselves up for more failures. Why bring so much pain in your life when there is so many

Good that happens every day? Why not start recognizing and celebrating all the good that happened that day? What gets rewarded, gets repeated. When you celebrate your victory often, you'll invite more amazing things into your life. And soon, you'll be conditioned to notice the Good in every little thing that happens in your life.

And soon, you'll develop powerful habits such as gratitude, consistency, tenacity, and self-discipline. So how do you celebrate? You don't have to complicate things.

Celebrating means giving yourself pleasure. It could be as simple as patting yourself on the back and say to yourself 'Good Job!', treat yourself a whole-hearty meal, giving thanks, listening to your favorite music, or meditating with gratitude.

The key is to give yourself pleasure. Often, we wait for others to praise, acknowledge, and reward us to feel happy, appreciated, and fulfilled. Why not be proactive and reward yourself when you can choose to be happy right now?

Conclusions First Part

Do not give up on your dreams. It might take time for you to achieve your goals, but with the power that is embedded in your mind, you can conquer the world one day at a time. Never let procrastination cripple your progress. Do not let failure deter you; learn whatever lesson that comes with failing, then restart your journey.

The more you use your skills you will discover other hidden treasures of ability that were buried in you. I can guarantee that you will be astonished at the things that you can do. Your talent can take you to places you have not even started to imagine.

The hard work and the sacrifices you make to attain success will be worth it. When you start your journey to success, each challenge you overcome will become a distant memory because the rewards will outweigh the struggles. Remember success is a lifestyle, one that a taste of it is very addictive. You can have your taste of success too. I might not know your name or where you are from.

I might not know what you have been through or what you are going through now. However, I know that you have something within you, something that sets you apart from the rest of the individuals in your line of work.

You need to find that part of your skill because it is your key to the door of success. No one in the world can do it for you, this is something you have to do on your own. I believe the fact that you have purchased this book is because that key, which is a part of your skill, has been nudging you. It is telling you that you can do this, you have what it takes. Do not be afraid. All it takes are some small steps, and the small steps will one day be a very long journey, a journey of no regrets because you tried and had achieved.

Everything in life is a choice; the only thing we do not have a choice about is when we will die. If you have created doubt within yourself at this moment, thinking that you do not have the money to invest in your plans, then the money is not hindering you from achieving your goals, only you are preventing yourself from progressing. Watch your inner thoughts; they will influence what you do.

Speak positive words to yourself every day. You will have many options, but never choose to quit, no matter how bad things get. Work on you, self-education is a powerful tool. Utilize technology.

Make some connections with people, who are successful in your field, sell them your plans. Describe your goals with such passion that they have to listen to you.

Prepare yourself for the "No's," it is just a part of your learning experience on your journey. Do not get mad at those who will reject your plans. You cannot blame them for not understanding the greatness that lies within you. Most of the individuals who will tell you, "No" now, will one day plead with you to join their team. Keep a firm hold on your spirituality; it can ground you, bring you clarity, and motivate you to understand your inner self.

Finding your place in the universe and with spirituality does not need to be religiously based. Spirituality is a state of consciousness that enables you to specify your desires and future with a clear mind and relaxed body.

So dear reader what will be your choice? I believe you will take your first small step to success, by conquering your fear and misery because you are going to build your success rituals as soon as you finish reading this book. Thank you for reading this book, and I am looking forward to seeing you at the top of the ladder of success. Good Luck!

- **Goals that never happen**

How many incomplete goals do you currently have on your agenda? If you're anything like the vast majority of us, then chances are that you have hundreds of projects that you started and never completed, countless goals that you told your friends but never saw through and all kinds of dreams that seem to be getting less and less likely to come to fruition.

And it's for this reason, that you may find people roll their eyes when you tell them your 'next big project'. When you start a new training program to lose weight and everyone – including you – knows that you're likely to have lost interest by month two.

Or when you talk about the app you intend to make, the website, or the business project. Or when you talk about that dream trip to Japan…

This is the way of things for many of us. We work incredibly hard at things we don't feel passionate about just to put food on the table but when it comes to fulfilling our dreams, we are remarkably ineffective.

It's time to change all that and to start making those goals happen. But how can you turn it all around?

How We're Going to Fix Your Goal Setting and Help You to Start Living the Life of Your Dreams

Accomplishing goals is about strategy, it is about making a cognitive shift to change the way you're thinking and it's about being smart about how you approach each goal. It's also about knowing how to choose your goals and even how to phrase them.

This book is going to show you how to make those changes then. You'll learn how to choose and write goals effectively, how to write effective action plans and how to make sure you stick with your goals and never give up.

But this book is going to be a little different than most goal-setting tomes, too. After we've given you the broad tools you need to start setting and accomplishing your goals, we're then going to take a look at how you can begin to put them into practice.

Because while a goal can be pretty much anything, for many of us they are going to fall into one of a few different categories. Most of us have goals for our relationships, goals for our fitness, goals for our careers and goals for travel. We're going to provide not only the abstract strategies you need to start making effective goals then but also the step-to-step processes that will let you apply these strategies in each of these areas. By the end of this book, you'll be adept at setting and accomplishing any goal. And at the same time, you'll have powerful strategies for improving your relationships, your fitness, your career and more.

Ready to change your life?

Chapter 8: The Most Powerful Skill you can learn: Setting Goals

Learning how to set goals properly is arguably the most powerful skill that you can learn. Why? Because it will allow you to then accomplish a huge range of *other* goals. When you know how to set goals, it allows you to effectively work toward anything. This is the key to unlocking pretty much everything you could want from life.

So ironically, the first goal you should focus on is the goal of setting goals! And until now, you've probably been doing it all wrong…

The Problem With Your Current Goals

How can a goal be wrong?

Sure, any objective is a worthwhile one, but the way that you phrase your goals and structure them is going to massively change your likelihood of finding success.

Let's take weight loss as an example because it's one of the more straightforward goals that is easiest to implement.

When you set out to lose weight, you should start with a concrete goal. And for most people this will look something like this:

"Lose 2 stone by next year" This is a terrible goal.

Why? First of all, it is far too vague. How are you losing weight? Weight from where? Why do you want to lose weight? What do you want to look like?

At the same time, it's out of your control. Even if you are completely committed to your goal, you may find that outside forces prevent you from being successful. Maybe you get ill, maybe you accidentally follow the wrong program, maybe it turns out you have a bad metabolism!

Finally, the goal is too far in the future. If your goal is to lose weight by next year, that then essentially gives you a license to procrastinate. The target is so far away, that you indulge yourself in a little overeating or put off exercise for a while and not worry about it until next month.

6 months pass and you realize you're *further* from your goal. And because it's too late, you're likely to just give up at this point.

Not a good goal!

- **What Good Goals Look Like**

So, what does a good goal look like? How might you phrase this same objective in a manner that will increase your chances of success?

The first thing to do is to focus on things that are immediately within your control and that are not influenced by outside factors at all. These goals should be things that you can accomplish in a guaranteed manner and that you will be immediately graded on a pass-fail basis.

So for instance, instead of aiming to lose 2 stone by next year, you would use this goal: "I will work out three times a week, every week, for at least 15 minutes"

Now *that* is a goal that you can aim for. Regardless of your metabolism, or of injury, or any other outside factor, this is a goal that you can accomplish. It also means you can't 'put off' the goal and it means you'll never reach that disappointing point where you can no longer stand any chance of completing it. At any point in your life, there is no reason that you can't set out to accomplish this goal and expect to be successful.

But by focussing on this small *short term* goal, you will then find that the long-term goal of losing weight takes care of itself.

- **How to Formulate Your Goals**

But that doesn't mean that any short-term goal that is binary is going to cut it.

First of all, you need to know what you want and make sure that the goal you are setting for yourself is going to help you get there. You need your goals to be intrinsically motivating and that means that you have to feel truly passionate about them. It's only been following a goal you feel excited about, that you will find you have the energy and motivation to keep going.

Working out for 15 minutes a day is an effective goal because it is sure to take you closer to your broader goal of losing weight. By keeping that end goal in mind, you should stand a better chance of staying motivated to work out even when you're feeling tired, or when you're feeling low on will- power.

And you shouldn't just be aiming to 'lose weight' either. Instead, you should have a more concrete vision of what this entails. Do you want to be thinner? Do you want to be more muscular? Why do you want that thing? Is it so that you will be more attractive to the opposite sex? Or because you want more energy? Be honest with yourself and listen to that drive inside that is pushing you toward the goal you want to accomplish.

If your goal is to make money, then try to focus on what the emotional hook is that is making you want that money. It likely boils down to more than cash – maybe it's status you want? Power? Confidence? Freedom? Only by really understanding the true nature of your dreams can you a) take the fastest route to accomplish them and b) maintain the drive and motivation you're going to need to get there.

This is going to require some soul-searching!

Moreover, you need to ensure that the goals are achievable and realistic and that you have broken them down into small enough steps. Case in point: our goal for weight loss is to work out 15 minutes per day. That's a tiny amount but it works because it's achievable and realistic. If you make your goal harder – such as working out for an hour a day – then you're going to find you're quickly disappointed when you can't find the time or the energy. You'll put off the exercise and make excuses. The best part about training for just 15 minutes is that once you *start*, you'll often find you go for longer.

Put it this way: it's much better to have a small, easy-to-accomplish goal and *stick with it* than it is to have a massive, life-changing goal that you can't manage!

But of course, if you're making your goal smaller, that means that it will take you longer to reach the eventual destination you're gunning for. This is not a problem: this is just another thing you need to accept if you want to accomplish anything. Things worth having take *time*. Take small steady steps and enjoy the journey.

Chapter 9: The formula: how to Structure the Goals and Implement your Plan

Now you know the basics of what makes a great goal, it's time to start building these kinds of goals for yourself. In this chapter, we'll lay out some simple instructions that you can follow to begin putting these ideas into practice.

Later we'll be applying this same formula to different areas of your life so that you can start going after a better body, a better salary and a better love life. But in each instance, we'll be reapplying this same strategy.

- **Step 1: Visualization**

The first and most important step is to visualize what you want and to *understand* what you want. We already discussed this a little concerning becoming richer. Often you'll find it's not the money that you want but rather what that money *represents* in terms of your lifestyle or your status.

The same goes for being fit. It's not enough to want to be thinner or healthier, you need to understand your motivations for wanting that. Do you want to feel more physically capable, perhaps become a professional athlete? Do you want to prevent the deterioration that many experiences as they age? Or do you want to look amazing so that you can be more successful with the opposite sex?

The best way to get an idea of what you want from life in any given area is often to just visualize your future. That means closing your eyes and just calling to mind your ideal future. Where are you? What do you look like? What do you do for a living? Who are you with?

By picturing your future in this abstract way, you'll be able to start analyzing what it is that you're trying to accomplish and from there you can begin to look at the more concrete steps you'd need to take to get there.

Some other strategies that can help with this are:

Looking at your role models and seeing what they have in common

Thinking about the things that excite you, your hobbies, the things you're a fan of, etc.

Thinking about the last time you felt truly happy, or truly alive

From there, it's also a good idea to think about the actual reality and to visualize what it would be like to get there and to live that life. Do you still want it?

For example, it's very easy to want to be a rock star in theory but you might not like the actual lifestyle: it would mean spending a lot of your life touring, being in the public eye and probably struggling to raise a family.

This is why we're thinking in abstractions at this point. Because you may find that the reality of being a rock star is not something you want – in which case you're going to start again and tap into what it was about that lifestyle that appealed to you. Are there other ways you can satisfy the same emotional goals? If you want to be recognized for your music, then you could try playing an instrument on YouTube or Band Camp! If you just want to be a professional musician, then you could compose music for computer games or videos.

But it doesn't all have to be about your career either: you could just as easily find that you're happy just busking, or making music in your spare time.

Getting to the core of what you won't like this can also help you to overcome impossible odds. If you want to be an astronaut for example, then you might have to come to terms with the fact that you are too old and it's now unlikely to ever happen. But ask yourself *why* that appeals to you on an emotional level. Maybe it comes down to your love of space, in which case you might be equally satisfied by being an astronomer? Maybe it comes down to your love of exploration and discovery, in which case you could be an explorer, or maybe just a researcher.

- **Step 2: Assess Your Situation Honestly and Thoroughly**

The next crucial step is to assess your current position versus the ideal one that you have visualized. This is where you're going to analyze the gulf between real life and your dream future and then try and find the best way to *bridge* that gulf is.

Making an honest appraisal of your current situation is a very important way to assess your current position and to thereby to get an idea of your strengths and weaknesses.

And in particular, you need to think about what advantages you have, what networks, what contacts and what opportunities. You may feel that you have none but that probably means you just haven't been thorough enough. As the saying goes: there's no such thing as a lack of resources, only a lack of resourcefulness.

This is also where you're going to analyze just how likely your goals are and then perhaps re-phrase them on that basis. If you've seen that you aren't likely to become an astronaut, then it's time to create a more achievable goal such as becoming an astronomer.

If your goal is to date incredibly hot women, then perhaps it's time to reassess and at least *start out*

by aiming for women that are on a similar level to yourself.

Your mantra for this step is to assess your situation honestly and then take the 'path of least resistance'. You're looking at the maximum benefit from the minimum time and work.

- **Step 3: Formulate a Plan**

This brings us to the next step, which is to formulate a plan based on your current situation, where you want to be and what options you have available to you.

For losing weight or getting into shape, this means looking – for example – at the different training programs. However, by making an honest assessment of yourself and your situation in the last step, you should be in a better position to choose a system that appeals to your particular strengths and weaknesses and that you are likely to see through.

So many people will pay for expensive training programs that involve eating a very strict diet and working out 10 times a week for an hour each session. But is that realistic? If you've tried to stick at previous workouts and have failed, then the answer is *probably not.*

When you assess your current situation, that also means assessing where things when wrong in the past and what your lifestyle and personality will allow for.

And by knowing this, you can then look for a training program or devise one that will work to *your* advantage. Maybe that means finding a way to fit CV in around your routine, or maybe it means sticking to a diet that you will find enjoyable and convenient.

The same goes for plans for travel and your career. It's time to get real and to get your head out of the clouds. Stop dreaming about traveling the world and instead, think about how you're going to travel more despite your responsibilities, budgetary limitations, etc. Stop wishing you were rich and start thinking about how you're going to climb the ladder in your career to get there.

When making your plan, it's also important to think outside the box and to reject the generally accepted beliefs regarding what you need to do to accomplish each goal.

- ***Reject the Norm***

Because we are only really taught one way to get what we want and that is to progress through our careers. And this is why so many of us get stuck. We decide we want to be rich and so we work harder, instead of realizing that we could be wealthier on our current salaries by spending less and perhaps finding a secondary income. We think the only way to become successful in music is to keep working our day job to pay for it. We think that the only way to travel more is to work harder and then retire early.

But the costs of living will inevitably go up to meet your salary, you will have less and less time as you work harder and harder and take on more responsibility and you'll find there's never 'a good time' to accomplish your goals.

And so instead you need to take the path less traveled. There are other ways to getting to where you want to be and if you're just banging your head against the wall, then it's time to rethink that strategy.

Nothing is stopping you from starting a business in your spare time *right now*. There's no reason you can't quit your job and start traveling *tomorrow*. You have the abilities you need to begin applying for higher-paid jobs. What's holding you back?

- **Step 4: Phrase Your Goals in Small Steps**

Now you know what it is you want to achieve and how it is you want to get there, you're going to hone in. You now know the 'bigger picture' and it's time to think about the small details instead.

You know you want to get fit, you know that going to the gym is not viable for you and you know that working out from home makes a lot more sense.

So all that's left to do is to phrase this as a goal that you can focus on every day or week. Hence: "I will work out for at least 15 minutes every day"

Maybe you've decided you're not so interested in toning muscle but want to start by focussing on losing weight so you'll look better in a suit and feel more energetic. In that case, your goal might be:

"I will walk to and from work every day that it isn't raining"

There's nothing wrong with having more than one goal or making more detailed goals either. You might couple this with a secondary goal, which could be:

"I will not eat anything on my 'foods to avoid list'"

Focus on these small steps and get yourself closer to your goal one bit at a time. Likewise, if you want to advance your career, then your goal might be to:

"Take every opportunity that arises to enhance my CV"

Or

"Apply for one job in the evening, three times a week"

Some of your broader goals are going to take multiple steps. For instance, if your goal is to become a famous musician, then perhaps you should take the following steps:

Learn to play the guitar by spending half an hour each evening, four days a week

Save $15 a day to invest in studio equipment

Output 1 video a week to build an audience

Output 2 videos a week to build an audience

Continue to output 2 videos a week and spend 1 hour per week in self-promotion activities

Spend 2 hours a week working on an album to sell from the channel

It's a long process but it's also a *real strategy*. It's a strategy that you need to succeed. It represents a cognitive shift where you're no longer daydreaming about being a famous rock star and you're instead looking at concrete, realistic, achievable steps.

And that's when you start making real, actual progress!

Chapter 10: Let Go of Fear

I'm going, to be honest with you now: there's a chance that you already know this deep down.

It makes logical sense that you should be making small, concrete steps to achieve your goals rather than making bombastic plans to 'become a rock star' or abstract visions like 'get richer'.

So what has been preventing you from doing that? Two things:

It's a lot of work – it's much easier and more satisfying to dream big and get the reward that comes from that, rather than face the reality of grinding towards your goals. We'll be talking about this later in the book when we discuss how to stay motivated and stick at your goals even when the going gets tough. Then there's the matter of feeling it's not the right time: you procrastinate instead of looking for other work. Again, this just needs a bit of rocket fuel, which we'll be looking at later on.

You're afraid. This is what I see so often and it's what condemns so many of us to a dull and unexciting lifestyle. We just don't want to take that leap and put ourselves out there. And in fact, it's easier to imagine ourselves as being very successful and to pretend we're going to get around to it than it is to put ourselves out there and risk having our ego shattered when things don't go our way.

We're about to address that second issue. Because if you want to be successful, then it's no good to continue procrastinating or trying to put off taking that plunge!

- **How to Know if You're Procrastinating?**

Some examples of procrastinating include:

Spending ages reading books and researching the topic instead of just getting stuck in. I see this a ton when it comes to fitness goals. So many people will spend countless hours reading books and blogs on fitness programs, hiring consultants and buying gym kits. But the one thing they never do? Start, working out! There's nothing wrong with researching health and fitness of course. It should be applauded. The problem is when you use this as a convenient excuse for not training. The reality is that *any* training program is better than nothing. If you want to start getting into shape – if you stand *any chance* of success – then you should start doing press-ups and pull-ups right now. There is simply no reason not to. You can then improve your routine over time but you start NOW.

Working on projects and never completing them. I work as an app developer and have released two highly successful apps in my life that have together earned me in the region of $90,000. Not life-changing amounts over a few years but certainly enough to make my life a little more comfortable, especially as they continue to earn money while I work my regular job. As a result of this, I am often approached by people who tell me that they're planning on releasing a successful app too. They then work on it for three years and never release it. The difference between them and me? I released my app when it was an MVP – minimum viable product. This is called the 'fail fast approach' and we'll talk about it later. Point is though, I put myself out there whereas they made excuses. Perfectionism is often just a delay tactic. Assess yourself!

Claiming the time isn't right. We touched on this briefly but just to recap: the time is *never* right. You're not traveling now because money isn't good? Sure, save up some cash – but by then you'll probably be at an exciting point in your career and not want to take a break. Then you'll have a partner and not want to leave them. Then you'll have a kid. There is never a good time to start a relationship, to get married, to have kids, to travel, to start a business. You do it anyway. And if you're worried what other people might say? Then follow the advice to 'ask for forgiveness, not permission'. Do it and worry about the consequences later. If it means that much to you, then it is the only option you have.

Ignoring your dissatisfaction. Do you know anyone in your life who wants to be in a relationship and who ignores this fact by throwing themselves into their career? Every

post on social media is about how excited they are about their new job or their travel. But you suspect that really, they just wish they had someone to go home to? In this case, they are trying to cover up one lacking area in their lives by focussing on the other. What about people who claim they are happy without pursuing their dream career because they have a family? Sure, that's great… but why not go for both? And that way be able to inspire your children with your inspiring story? Don't make this mistake because you need to be fulfilled in *every* area of your life if you're going to be truly happy.

- **Fear Setting**

If you still can't overcome these psychological blocks though, then its time to employ a technique known as 'fear setting' that was described by Tim Ferriss in his book: *The Four Hour Workweek.*

The idea here is simple: you are going to write down all of the things holding you back and all of the things you're afraid of and then you're going to present counterarguments, contingency plans and more to *remove* those fears.

So take a moment to think about your goals and dreams and then write down all of the things that you want to accomplish. Write those goals and the steps you need to take as we discussed in the last chapter and then think about taking that first step right now. What's holding you back? What are your fears? Be honest and thorough and make sure to include every possible concern.

Let's say you want to start your own business. Here are your fears and concerns:

You don't have the money

Taking out a loan may be reckless and leave you in serious debt if the business isn't a success

Your partner might see your investment as irresponsible and lead to relationship problems

You might lose your job and find yourself without a stable income

You might be unable to find future work and that could lead to your family going hungry and you losing your home

Your business might be a failure and make you look like a failure too.

Now go through each of these objections and address how likely they are and how you can deal with them/prevent them.

For example:

You don't have the money and taking out a loan may be reckless and leave you in serious debt if the business isn't a success,

Consider a PayPal loan, this is a loan that you pay back only through PayPal income, meaning that you won't owe anything until you start earning try Kickstarter,

bootstrap your business – design it in a way that will allow you to start the business for less.

Consider asking parents for a business loan.

Look for a business partner with capital to invest

Your partner might see your investment as irresponsible and lead to relationship problems.

Your partner is more likely to support you in your ambitions

If you use the above methods, you can demonstrate that you have been sensible and taken every precaution

You can even take out business insurance,

your partner might even be able to help you bring in extra income to support your goals

Have a rainy day fund

Explain to them the risks and why it's important to you

You might lose your job and find yourself without stable income &

You might be unable to find future work and that could lead to your family going hungry and you losing your home

In most cases, you'll find that your employer will offer you your job back if you need it

At the very least, you can probably find lower-level work to fund your survival

Even if that means just doing a part-time job

You don't have to quit your day job until you've proven to yourself that you can make money from your business idea

Or even maintain a part-time salary in the meantime

You can probably survive on a lower salary than you think and for longer than you think

Your business might be a failure and make you look like a failure too

You will do market research and take every precaution to ensure your plan is a success

You will gain advice from knowledgeable third parties

Who cares what other people think

The alternative – never trying to make anything of yourself or pursue your passions

– is far worse

Okay and with that out the way, now we can start making progress in the various areas of your life that you want to improve!

Chapter 11: How to Achieve your Fitness Goals

We've seen the basics of how to accomplish your general goals, now it's time to accomplish *specific* goals. For this chapter, we're going to look at fitness and how you're going to apply the principles we've discussed to get into awesome shape.

So the first thing you need to know is why you want to improve your fitness and what you want that to look and feel like. Is your goal to get fitter so you can play sports again? Do you want to look awesome for your satisfaction? Do you want to be powerful so that you feel more physically intimidating? Do you want to be healthier? Or maybe attract members of the opposite sex?

And what is your current situation? What have you tried in the past? Why has it not worked? What is your current shape and size? What are your physical strengths and best attributes? What do you enjoy doing? How much time do you have?

This is all very important because it is going to drastically change the way you go about accomplishing your objectives.

For example, if you are a man and your goal is to be more physically intimidating, then you might decide that it makes the most sense to bulk. This means adding the most mass possible in the shortest amount of time, to become a tank. It involves eating a ton of calories and even more protein, resting a lot and lifting heavyweights.

On the other hand, if you want to become toned and lean-to attract women, then you are going to want to eat less and get more aerobic exercise such as walking, running, skipping, etc.

You also need to think about the exercise that you enjoy doing, the exercise that is practical to work into your routine, any physical limitations such as illnesses or joint problems, etc.

- **How to Set and Stick to Realistic Goals**

One of the most important considerations when coming up with a training program is making it fit into your routine. Think about when you have time free, how your energy levels are at different points during the day and what you can do to capitalize on the moments in your routine that are free for training etc.

- **Fitting it In**

One of the best ways to lose weight, for example, is to walk more. Walking is ideal because it burns a good number of calories without exhausting you, or making you sweaty. That means you can conveniently fit it into your routine and do it regularly without it becoming unfeasible.

And most of us can easily fit more walking into our routine. For example, you might find that you can use your lunch break at work to go on a long walk. If you have 60 minutes at lunch, you can eat for 10 minutes and spend the other 50 walkings (it's best to walk at the *end* of the 60 minutes). A 50-minute walk each day should easily be enough to hit your 10,000 step goal, which is around 5 miles and should lead to an additional 3,000 calories (roughly) burned each week. That's the number of calories you normally burn in a day. More importantly, it will build your fitness significantly, give you more sunlight and fresh air, etc.

So forget trying to do intense HIIT workouts 5 times a week that leave you exhausted… just go for a nice walk that will conveniently fit into your routine!

Likewise, you can fit a walk-in by getting off the bus early, by walking home from work, etc.

The same goes for diet. I always advise clients to stick to a rigid diet *only* in the morning and at lunch. Why? Because most of us will want to make our evenings a time to enjoy a fun meal with our partners. Or we want to go out with friends and enjoy pudding. Conversely, breakfast and lunch tend to be more functional – eaten alone and in a hurry. That means you can much more easily reduce your calories or your carbs at this time during the day and then 'cut loose' in the evening.

Think about ways you can make this more convenient for you too. If you pass a shop that sells protein shakes in bottles each morning, then maybe switch your morning coffee for a morning protein shake. This is ideal if you find that the thought of mixing your protein shake and getting it all over the floor potentially is putting you off of actually eating it!

Another example might be to workout from home if you are struggling to get to a gym or to take up swimming if there just so happens to be a pool next to your office.

- **Enjoy It**

Your exercise should be something you enjoy. If you have tried and failed to build lean muscle with weights, then clearly you're not cut out for it. It just doesn't appeal to what you enjoy.

But all of us should find there's *some* form of exercise we enjoy. Maybe you should get yourself a pair of parallel bars (which are very cheap) and take up gymnastics or hand-balancing at home?

Or instead, how about taking up rock climbing. Rock climbing is *fantastic* for building big, powerful muscle, particularly in the lats and forearms. Maybe you'll find you love boxing: getting yourself a heavy bag is a great, enjoyable way to build big shoulders in particular. Or maybe you might be cut out for powerlifting?

Whatever the case, find a form of training. This is what all the most powerful people with the most incredible physiques have in common. They don't just love being big, they love *getting* big. They eat, sleep and dream the gym and they love everything from the feeling of the chalk on their hands to hang out with other swole people.

You need to discover that passion not just for the end destination but for the journey to get there.

- **Play to Your Strengths**

Some people are ectomorphs naturally, some are endomorphs. This determines whether you're a big, bulky type or a lean 'hard gainer'.

Where possible, try to align your goals to your natural strengths (remember step 2?). So for example, if you are an endomorph, then you can focus on becoming a massively powerful hulk. If you're an ectomorph, then why not go for the lean look that a lot of people love?

There's nothing wrong with chasing after the harder dream of course but if you're flexible, shoot for the one that you're already gifted in. That way, the results will come faster and you'll find it more intrinsically rewarding, more quickly.

Another tip is to find role models that are similar to you. Look for people who started in your situation, people who have body types similar to your own, but who have made the very most of them. Those are the people to listen to when it comes to training advice because they have worked with (most likely) a similar genetic starting point and a similar set of circumstances in life to begin with.

- **Take it Slow**

Remember what we said in the earlier chapters: a good goal for fitness should involve working out for 15 minutes, maybe even 10 minutes. Don't come up with insane strategies that involve training twice a day, or you'll find that you gain muscle quickly and lose it quickly. Be willing to see small improvements over time so you don't burn out.

Conversely, though, don't take it *so* slowly as to not see results. The objective here is to use the MED – or 'Minimum Effective Dose'. That means you're committing just enough time to see progress, so that you can start to assess and judge your strategy and so that you can improve it over time. Don't do more, don't do less.

By doing all of this, you should have come up with a training program that is effective for you specifically and for your lifestyle and genetics.

If you have tried and failed to take up weightlifting several times in the past, then maybe it's time that you took a different approach by swimming three times a week after work. Or by getting a heavy bag and punching that for 40 minutes a few times a week. Maybe you just do 15 minutes of press-ups before bed.

Whatever the case, start doing something right away and then experiment to find what works for you.

Chapter 12: How to Achieve Your Career Goals

Too many people have mistaken ideas when it comes to their approach to their careers. We often believe that working incredibly hard in jobs that we don't truly enjoy is 'responsible' and what adults should do. We often feel that we don't have any choice when it comes to what we do for a living. We often feel scared to try anything else.

And this is why so many of us are unhappy in our careers: we just 'let them happen' and accept the career path that we fall into. We leave school or college, take the first job opportunity that comes out way, and then work hard to progress up the ladder. We never take a moment to ask: is this what I want? Do I have a choice?

Here are some ways to apply the principles that we've discussed in making progress in your career...

- **Knowing What You Want**

The first and most important thing to focus on here is step 3 – coming up with your plan. It's time to acknowledge that you don't *have* to continue working a job you don't like and there's no reason that you even need to focus on your career at all.

The first myth we need to dispel then is the notion that you need to get your sense of satisfaction and progress from your career at all. That is to say that you should be able to get the same satisfaction as a hobby. We often feel that our sense of self-worth and achievement is tied up in our careers and that we need to work harder and harder to feel like we're progressing in

life. But while you might be CEO of a logistics company, you are still ultimately in charge of making sure people get staplers – when your passion might be painting works of art.

This is why you can often do better to simply switch your focus to your 'extra-curricular activities'. My sister did this as an artist when she realized that the reality of her intended field (creating props for movies) was not quite as idealistic as she had hoped. So instead, she took on a job that would pay the bills by working as a saleswoman and then used her *spare time* to work on her creations in her own time.

She has gone on to receive quite a following on social media and has sold several of her works to private buyers. So although her career isn't something she gets particularly excited about, she still gets that sense of progress and excitement and doesn't *need* to keep taking on more responsibilities to feel happy and fulfilled.

And that also brings us to the other point: your wealth isn't entirely determined by your career either. You can just as easily augment your income through other means – whether that means renting out your room or whether it means cutting the neighbors' hair.

This is once again why it is important to consider the precise nature of your goals – if your goal is to be richer, then you can do it by shrinking expenses, by finding other sources of income, etc. If your goal is to get more status then you may be happy to progress in your current career. If it is to be fulfilled in your artist endeavors or to be acknowledged for your ideas, then you may prefer to focus on working on projects outside the office.

This is essentially what we refer to as 'lifestyle design'. Lifestyle design means that you're focussing on what you can do to create your perfect lifestyle and you're looking for the path of least resistance to get there. This might not mean working more – it might mean working less and even taking on a 'menial' job so that you can put more energy into other areas of your life.

Heck, it might mean creating income from elsewhere so that you can afford to work 4 days a week. Why not!

- **Creating a Fool-Proof Strategy**

We've addressed the power that fear can have over us and the way it can prevent us from going after our goals. This is especially true when it comes to achieving things in our careers. And for that reason, it makes sense for us to take a look at some of the things we can do to make our career goals less risky.

For example, a lot of people will make the statement that they want to look for another job but that they can't because they have too many responsibilities. They might even make the unfounded claim that they wouldn't be able to find another that would pay the same salary... without looking!

But there is no reason that this needs to be seen as a risky undertaking and no reason that you should be afraid to look for work: the simple answer is just to look for other jobs while you are working your *current* job. Spend a couple of evenings looking at other jobs and applying for them and only leave your current job when you have a new one: zero risks.

The same goes for starting a part-time business. You don't have to immediately transition from one job to another when you can simply use your spare time in the evenings or on the weekends to work on your new business idea. Only once you are certain it works should you then consider leaving your current job to take on the new one and this will present you with another risk-free way to transition to a job or career you love.

You can even try reducing your work hours and then use that extra free time to work on your business. Take a part-time job and during your longer free hours, work on your business project.

The same goes for investment. If you need investment to create a business idea then there are lots of risk-free ways to get it. Using Kickstarter these days is a great option for example and involves zero risks as well as a great way to test the reception for your idea.

Likewise, you could ask your parents for investment, you could get a business partner friend, or you could take out a credit card loan. As long as you don't quit your current job, you can just make sure that the loan repayment terms are something you could manage to pay off if you had to and that way, you won't be putting yourself at any risk.

If you want to make this happen, then you can always find away.

And if you have assessed your vision and you just want to be a rock musician, then don't be distracted by trying to get rich. Focus to start with on doing the thing you love and finding more time for it. Let success come as a by-product.

As soon as you start working on your project, you'll find that it is rewarding and you now have a drive and passion that wakes you up in the morning and makes you more animated, more passionate and more exciting to be around for others even. It doesn't even *matter* if you are a success. And that's why you should also view 'failure' simply as a chance to reassess your strategy and try something else. When you take this approach, there is no way you can fail.

- **The Path of Least Resistance**

Remember, genuinely going after something you want means taking the most direct and practical route to get there: the path of least resistance. In this case, that means creating a business idea that you can realistically accomplish, or designing one around your current contacts and ideas.

One common mistake that a lot of people make is coming up with ideas they think will change the world. If that is your vision, then it doesn't also need to be a money-making venture, to begin with.

But if your vision (step 1) was to become wealthy, maybe to gain financial independence, then the most effective way to accomplish that goal is to focus on tried and tested methods for making money.

That is to say that you don't need to break the mold and come up with whole new business models. You don't need to become the next Mark Zuckerberg. Because there are billions of these huge projects that fail every year.

Meanwhile, just count the number of successful shops, clotheslines, resellers, building companies, hairdressers. There's nothing wrong with taking an idea that you've seen work and then just following it through to the letter. You now have a blueprint for success and you aren't having to reinvent the wheel.

Likewise, think about your resources and contacts. If you happen to know the editor of a gardening magazine, then that is an incredibly powerful contact to have and you should make the most of that

start a gardening service and use them to advertise!

You should also play to your strengths and if you know a lot about gardening, then once again, this is a good choice for your career. Very often, branching out (no pun intended) to start your own business will make the most sense if you stick in your current industry: this way you'll have the expertise, experience, and contacts to give yourself a great head-start.

Remember step 2: assessing your current situation and your resources. Make a list of everything you have available to you, all your skills, all your limitations and then think about what business and lifestyle changes will help you to get.

- **The Fail Fast Model**

Remember when we discussed how fear could hold some people back and one way this presented itself was when someone would work on perfecting their product without actually *releasing* anything? Not only is this a blatant delay tactic but it also means that if you eventually *do* release your product, you risk suffering a devastating defeat if it doesn't go to plan. This happened to a

friend of mine who had an idea for a business and then spent the next 3 years perfecting it. He trademarked the business name, took on a legal advisor, even paid for an expensive launch party! All for what essentially amounted to a website. He tested the site in every browser and every display size *meticulously*, he conducted copious amounts of market research and he paid for tons of server space and bandwidth ready to cope with the inevitable huge amounts of traffic. But his upfront and ongoing costs were so high that he went bankrupt almost immediately.

The opposite approach is the 'fail fast model'. If you have an idea for a business, then you should create an MVP or 'minimal viable product'. This is the most basic, affordable and easy version of your product or service that you can release to the market immediately. That way, you can now test market response to it without having invested lots of your time and money into it. You throw lots of ideas at the wall, quickly putting together something that works. If the idea is successful, you can then invest time and money into it. If it isn't, you iterate, learn from your mistakes and move on!

Ù

Chapter 13: How to Achieve Your Relationship Goals

Relationships are something we often don't think of as 'goals' but they are in precisely the same way as any other. Maybe you're single and want to be in a relationship? Maybe you're in a relationship and want to make it better? Maybe you just want more success with the opposite sex? These are all worthwhile goals and all of them can be subjected to the precise same strategy that we looked at before.

- **Taking Stock**

Here though, perhaps the most important aspect to look at is step 2: an appraisal. You need to really take the time to assess the current state of your relationships and yourself and then to work on moving forward and improving those areas of your life.

This starts by looking honestly at your current relationship. A lot of people will remain in unhappy relationships because they can't face admitting to themselves that things aren't perfect; perhaps because they have a child or house together, perhaps because they love their partner.

But note that improving your relationships doesn't necessarily have to mean ending your relationship. You can work on a relationship just like you can work on a car or a business. You can improve the way the relationship works, improve your happiness in your role and generally see positive change over time. Would you be happier if you had more sex? Are you getting enough time to spend with your partner properly? Do you argue more than you'd like? Sometimes, it is just a case of making some simple changes which can help you to improve on those areas and your relationship will be better for it. Don't live in denial.

Likewise, if you currently aren't having any success with approaching people, or if you're single and you don't want to be, then you maybe need to address certain aspects of your game to change how you are coming across. This is a skill that can be learned like any other and often it comes down to appearing confident and presenting yourself well. If you can do that – without coming across as arrogant – then you will have much better luck approaching people.

Often people who never have any success in dating are portraying themselves in the wrong way. Maybe you're too shy to approach women/men and this means that you never get to choose who you date. Maybe it's a confidence issue and you feel that you *can't* approach them without being rejected.

Or perhaps you approach but you are coming across sleazy, awkward, or generally unattractive.

- **Creating a Plan for Dating**

The aim is to look confident, successful and likable. This sends a powerful and strong signal that others respond to as meaning that you're likely to be a great 'catch'. In other words, if you are projecting yourself as being very confident, then others will assume you have good *reason to be*. This speaks to our evolutionary imperative: to go after people who are of a higher status than us, people who will provide good genetic material for our offspring, or people who have resources.

And this is where we can employ step 3: creating a 'foolproof' and 'non-conventional' plan to achieve that. For example:

Head to a bar with some friends, chat with them and have some drinks. Try to appear like a fun group to spend time with.

While you're there scout out the room. Look for women/men who are within your reach. In other words, look for people who aren't *too* far out of your league. But don't be afraid to punch above your weight.

If you see someone you like, shoot them a smile. If they are at all interested, they'll smile back.

This *immediately* helps you to remove any risk of being rejected. If they're not interested, they won't look back and you'll just avert your attention elsewhere. If the person you spot is keen, they might even take the initiative and head over!

If not, you head over to them. Now crucially, don't focus on one the person you like but rather introduce yourself to the group and let your two groups mingle. This is your chance to demonstrate that you are fun, outgoing likable and confident. You are displaying alpha male/female behavior and making yourself the center of attention. What's more, is that

you're demonstrating that you are popular (as you have a group of friends) and that you get on with *their* friends. And what you're also doing, is showing that you might not be interested in them at all – which makes you much *more* desirable.

If things go well, then you can offer the person you're interested in a drink. This sends a clear signal without saying as much and if they come with you, that's your chance to get them alone. If they said yes to the drink, then you can ask them to dance – again without losing face if they say no.

And if This plan relies on a couple more factors, all of which are within your control. You need a group of friends to help you for instance and you need the confidence to approach the group and to make yourself the center of attention. This takes practice and that can be your short term goal – practicing this technique until it comes as second nature and you *do* feel confident and charming around members of the opposite sex.

- **Knowing What You Want**

Again though, it is important to take into account step 1 here: knowing precisely what it is that you want. It is crucial that you are approaching the right type of partner and that you are sending the right signal. It is a very different set of tools and approach that you need if you want to approach people for one night stands, versus approaching people that you want to have a long-term relationship with. The former means going to bars and looking for people sending certain signals. The latter might mean getting closer to a friend, or looking for someone who has a lot in common with you and who is *also* ready to settle down. If you want to play the field then you might consider using sites like Tinder.

Chapter 14: How to Achieve Your Travel Goals

What about travel goals? What if your goal is to see the world? Again, we apply our steps which means we look at the kind of travel that we want to accomplish then work out a way to make it happen that is feasible considering our specific circumstances.

So we start again with visualization: picture the type of traveling you want to do, know what it is you want to get from your traveling and think about the different ways you can accomplish those broader goals.

Then look at your circumstances. What is holding you back? Budgetary constraints? Family responsibilities? Fear?

Then make your plan based on this information and break it down into small steps. Again, this might mean thinking outside the box and taking the 'nonobvious' route to success. You don't necessarily have to take the obvious route by taking a gap year and traveling to various far-flung reaches of the globe.

- **Alternative Travel Strategies**

Perhaps you don't have the time or budget for that and would get just as much from traveling more locally? There are some incredible things to do see and do in the US if you're in America, or if you're in Europe then you have the whole of the

EU on your doorstep. This can present just as much adventure and variety and even if it's not exactly what you initially thought, it's still going to scratch that itch and that need for exploration and discovery.

Or how about just going for a shorter time? You can have a truly life-changing experience in just 3 or even 2 months. And you're much more likely to get a sabbatical lasting that long and be able to save the money. You can even change your strategy entirely and try taking lots of very small trips throughout the year. This might also be easier to convince a partner of, versus going traveling for months at a time.

Moneywise, you might be surprised at how little you need to travel if you go during non-peek seasons, if you stay on people's couches or if you use Air BnB. That means that you can earn a little money online to fund your travels.

Or how about asking your current job if you can be sent abroad? If the business has branches all around the world, this may very well be viable. Likewise, there might be a role that involves travel – or you could just apply for a job that involves travel. That way, you travel *while* earning money and have a good explanation for your other half.

And there's no reason you can't take your partner with you either, of course.

Conclusion Second Part

Do not give up on your dreams. It might take time for you to achieve your goals, but with the power that is embedded in your mind, you can conquer the world one day at a time. Never let procrastination cripple your progress. Do not let failure deter you; learn whatever lesson that comes with failing, then restart your journey.

The more you use your skills you will discover other hidden treasures of ability that were buried in you. I can guarantee that you will be astonished at the things that you can do. Your talent can take you to places you have not even started to imagine. The hard work and the sacrifices you make to attain success will be worth it. When you start your journey to success, each challenge you overcome will become a distant memory because the rewards will outweigh the struggles.

Remember success is a lifestyle, one that a taste of it is very addictive. You can have your taste of success too. I might not know your name or where you are from. I might not know what you have been through or what you are going through now. However, I know that you have something within you, something that sets you apart from the rest of the individuals in.

There are many more types of goals that you might choose to pursue and where you might choose to use this formula. For example, you might have goals that pertain to your finances alone, maybe to your property, maybe to your social life… perhaps your goal is purely to learn a particular hobby or to improve the way you dress.

The whole point of this system is that it can be applied anywhere and when you do that, it will help you to understand what it is you want and to make those aims concrete and tangible. This takes them from being dreams that you end up putting off forever and turns them into a series of steps you can use to make that happen. Sometimes this might mean reassessing your goals to make them that bit more achievable but if you're smart about this, they won't be any less rewarding. Maybe you can't be the next Brad Pitt or Angelina Jolie but there's no reason you can't start playing bit parts in movies if you think about how to structure your life around auditions.

It's about knowing what you want and then assessing the quickest way to get as close to that ideal as possible. And as soon as you start trying, life becomes a whole lot more rewarding and amazing. It's time to stop dreaming and start doing…

It's time to make it happen!

Book Bonus!

- **How to do more in a fraction of the time Word**

The pace and intensity of our lives, both at work and at home, leave several of us feeling like a person riding a frantically galloping horse. Our day-to-day incessant busyness — too much to do and not enough time; the pressure to produce and check off items on our to-do list by each day's end — seems to decide the direction and quality of our existence for us.

However, if we approach our days in another way, we can consciously change this out of control scheme. It just requires the courage to do less. It may seem simple, but doing less can be very difficult. Too often we mistakenly believe that doing less makes us lazy and results in a lack of productivity. Instead of doing less, it helps us enjoy what we do. We learn to do less of what is foreign and engage in less self-injurious behavior, so we create a rich life that we feel really good about.

 We learn to do less of what is extraneous and engage in fewer self-defeating behaviors, so we create a rich life that we truly feel great about.

Just doing less for its own sake can be easy, startling, and transformative. Imagine having a real and unhurried conversation in the middle of an unforgiving workday with somebody you care about. Imagine completing one discrete task at a time and feeling calm and happy about it. In this book, you will see a new approach.

The approach is equally useful for our personal life and our work life. The two hemispheres of our work and personal lives constantly reflect on and affect one another, each changing and/or reinforcing the other.

Every life has awesome meaning, but the fog of constant activity and plain bad habits can often obscure the meaning of our own.

Acknowledge and change these, and we can again enjoy the ways we contribute to the workplace, enjoy the sweetness of our lives, and share openly and generously with the ones we love. Less busyness leads to appreciating the sacredness of life. Doing less leads to more love, more effectiveness and internal calmness, and a greater ability to accomplish more of what matters most to us.

Chapter 15: Priorities

Not everything in life can be a priority. Many important things will compete for attention over your lifetime, but there are not enough hours in anybody's lifetime to give attention to everything that could potentially be a priority.

Determining your basic priorities is a key exercise in moving toward more efficient use of your time. Your basic priorities provide a means for making time choices, helping you decide where it is important to invest yourself and where you can let go.

- **Prioritizing**

Setting priorities is a matter of deciding what is very important. In this case, "important" means significant to you. What activities and roles give your life meaning? These are the components of your life where you would like to succeed the most.

Not everything in your life can be a priority. Many important things will compete for attention over your lifetime, but there are not enough hours in anybody's lifetime to give attention to everything that could potentially be a priority. Determining your basic priorities is a key exercise in moving toward more efficient use of your time.

Your basic priorities provide a means for making time choices, helping you decide where it is important to invest yourself and where you can let go.

Daily, you also have to learn to set task priorities. Prioritizing tasks includes two steps:

Recognizing what needs to be done

Deciding on the order in which to do the tasks

How do you determine what work needs to be done? For the most part, it relates to your basic priorities. To be efficient in your time use, you have to weed out the work that does not fit with your basic priorities. Learn to say "no" to jobs that look interesting and may even provide a secure sense of accomplishment but do not fit with your basic priorities.

You also have to be able to separate the tasks that require busywork that tends to eat away at your time. Many tasks that fill your day may not need doing at all or could be done less frequently. Task prioritizing means working on the most significant tasks first regardless of how tempted you are to less significant tasks out of the way.

Certain skills help in using time effectively. Most of these skills are mental. While it is not necessary to develop all of the skills, each contributes to your ability to direct time usage.

Time sense is the skill of estimating how long a task will take to accomplish. A good sense of time will help you be more realistic in planning your activities. It helps prevent the frustration of never having quite enough time to accomplish tasks.

To increase your time sense, begin by making mental notes of how long it takes to do certain routine tasks like getting ready in the morning, running a load of laundry or delivering your child across town to baseball practice.

Goal setting is the skill of deciding where you want to be at the end of a specific time. Goal setting gives direction to your morning, your day, your week and your lifetime. The exercise on deciding your lifetime priorities is a form of goal setting. Learn to write down your goals.

If you are like most people, goals are just wishes until you write them down. Keep your goals specific, as in "weed the flower beds in front of the house" rather than "work on the yard." Keep your goals realistic or you will continually be frustrated by a sense of failure.

Standard shifting is adjusting your standards as circumstances change. Your standards are what you use to judge whether something is good enough, clean enough, pretty enough, done well enough.

Perfectionists have very high, rigid standards, and they have trouble adjusting to the changing demands or circumstances of their life.

Develop the ability to shift standards so you can be satisfied with less than perfect when your time demands are high, instead of feeling as if you are somehow falling short.

Time planning is outlining ahead of time the work you need to be done in a specific period. Sometimes time planning is as simple as writing out a "To Do" list to ease your mind from holding on to too much detail.

At particularly stressful times, the "To Do" list may expand to include a more specific calendar of when tasks will be done. While a detailed schedule can be too confining to use all of the time, it is a good way to take the pressure off at exceptionally demanding times.

Recognizing procrastination is a skill in itself because procrastinators can do an incredible job of hiding their procrastination from themselves. Procrastination is needlessly postponing decisions or actions.

You might disguise the procrastination response with an excuse like waiting for inspiration or needing a large block of time to concentrate with your full attention, or need more information before tackling a project.

It takes skill to differentiate between procrastination excuses and legitimate reasons for delaying a decision or action. Without the ability to recognize when you are, procrastinating there is little chance of overcoming this immobilizing habit.

Chapter 16: Tips to Help You Set Priorities

Here are some tips to help you prioritize. It is important to use these tips regularly to help remain focused. Each of these techniques can help you in getting closer to your goal of becoming more effective with your time.

- **Tips**

Each of these techniques can help you in getting closer to your goal of becoming more effective with your time:

- **Assume ownership of your time**

Most individuals would be surprised if somebody reached in their wallet without asking and helped themselves to the money found there. But how different is that from letting other people help themselves to your time? Take possession of your own time and do not allow other people to make commitments of your time without your permission. It is not selfish to keep other people from consuming your time. Give your time freely when you want but do not make the mistake of undervaluing this resource, or feeling guilty when you do not allow other people to waste it. Think of time lately when somebody wasted your time. How could you have dealt with the situation better?

- **Prioritize**

Continually check yourself to see that you're working on the most significant things. Helping your child talk through a problem, he/she is having or discussing the day's events with a spouse or friend may be more significant than getting the dishes done or a load of laundry completed. Do not think of priorities only as tasks that need doing. As you remind yourself to direct yourself to the most important tasks first, you will find yourself letting go of tasks that did not need to be done in the first place.

- **Learn to say "no"**

It is not that saying the word is so difficult. It's more the feeling of guilt that many women experience as soon as they use the word. Try centering on the significant things that will be done because you used that two-letter word to decline something which was not a part of your priorities. Considering your past week, what are some things you should have said "no" to?

- **Protect your blocks**

Think of your day as numerous large blocks of time with the blocks divided by natural interruptions. Where you have control, keep your blocks whole, scheduling appointments and meetings, running errands at the beginning or end of a block instead of in the middle. Having an appointment in the middle of a block leaves little time at either end to tackle a major piece of work. Keeping your blocks of time as big as possible gives you a feeling of having more time that is available.

- **Delegate**

There is that "D" word. Delegating means assigning the responsibility for a task to somebody else. That signifies you no longer have to do the task, nor do you have to remind somebody else to do it. Being able to delegate some tasks is a way of freeing up some of your time for the jobs that only you can do. As somebody else learns to do a job, do not be tempted to take over if they are not doing it quite right. You have to learn that "done" maybe "good enough."

- **Think in terms of buying time**

There's an intimate relationship between time and money, where one can often be substituted for the other. The more hectic your schedule, the more reasonable it is to buy time by selecting goods and services that save you from investing time. Paying somebody to mow your yard or transport your kids to baseball practice are examples of purchasing time. What are some of the additional ways you can or do buy some time?

- **Learn to work with your biological clock**

People have a peak time of day when their energy is at its highest and concentration at its best. Determine which time of day is your peak performance time and plan your work accordingly. Keep meetings and routine tasks for other parts of the day when you have the choice. What part of the day is best for you to do a task that takes real concentration?

- **Break down big jobs into manageable pieces**

One of the sources of procrastination is that some tasks can seem too overwhelming to even begin. Learn to break down a large task into manageable pieces and then begin with a piece you know you can handle. The most challenging step in major undertakings is often the first one. Besides, you will have a greater sense of satisfaction as you complete each portion of the task and this can keep you motivated to the end. Think of a major task you have ahead of you.

How could you break it down into manageable pieces?

- **Work on overcoming procrastination**

Once you recognize that you are procrastinating, the next step is to begin overcoming this time-wasting habit. Besides, procrastination is a habit, a habitual way of dealing with tasks you find distasteful or that make you fearful of failure. When you see that you are procrastinating, make an appointment with yourself to take the first step toward completing the task. Determine exactly what that first step will be and then set a specific time shortly to begin the work.

- **Reward yourself**

Celebrate when a major task is completed or a major challenge is met. One of the problems with a hectic life is that you can be so busy that you fail to notice the completion of a major piece of work. You just move on to the next job without celebrating your previous success.

This failure leads to focusing on what is still left undone instead of enjoying what has already been accomplished. Set up a reward system for yourself that serves as both a motivator to get certain difficult tasks done and an acknowledgment that you are making effective use of your time. Be it a bubble bath, two chapters in your new book, or a phone call to a friend, acknowledge your accomplishment by rewarding yourself.

Chapter 17: Beat Procrastination

If you have found yourself putting off important tasks repeatedly, you're not alone. Many people procrastinate to some degree. The key to controlling this destructive habit is to recognize when you begin procrastinating, understand why it happens, and take active steps to manage your time better. In a nutshell, you procrastinate when you put off things that you should be focusing on right now.

- **Beating Procrastination**

If you've found yourself putting off important tasks over and over again, you're not alone. Many people procrastinate to some degree. The key to controlling this destructive habit is to recognize when you begin procrastinating, understand why it happens, and take active steps to manage your time better. In a nutshell, you procrastinate when you put off things that you should be focusing on right now.

- **How to Overcome Procrastination**

Follow these steps to deal with and control procrastination:

- **Step 1: Recognize that you are Procrastinating**

If you are honest with yourself, you probably know when you are procrastinating. Here are some useful indicators that will help you know when you are procrastinating:

Filling your day with low priority tasks from your To-Do List.

Reading e-mails several times without starting work on them or deciding what you are going to do with them.

Sitting down to start a high-priority task, and almost instantly going off to make a cup of coffee.

Leaving an item on your To-Do list for a while, even though you know it is important.

Regularly saying "Yes" to unimportant tasks that other people ask you to do, and filling your time with these rather than getting on with the important tasks already on your list.

Waiting for the "right mood" or the "right time" to tackle the important task.

- **Step 2: Adopt Anti-Procrastination Strategies**

Procrastination is a habit – a deeply ingrained pattern of behavior. That means that you will not just break it overnight.

Habits only stop being habits when you have persistently stopped practicing them, so use as many approaches as possible to maximize your chances of beating procrastination. Some tips will work better for some people than for others, and some tasks than others. Sometime you may just need a fresh approach to beat procrastination.

These general tips will help motivate you to get moving:

Make up your rewards. For instance, promise yourself a piece of tasty flapjack at lunchtime if you have completed a certain task. Besides, make sure you notice how good it feels to finish things!

Ask someone else to check up on you. Peer pressure works! This is the principle behind slimming and other self-help groups, and it is widely recognized as a highly effective approach.

Identify the unpleasant consequences of NOT doing the task.

Work out the cost of your time to your employer. As your employers are paying you to do the things that they think are important, you're not delivering value for money if you're not doing those things. Shame yourself into getting going!

Aim to "eat an elephant beetle" first thing, every day!

Remember: the longer you can spend without procrastinating, the greater your chances of breaking this destructive habit for good!

Chapter 18: Tips to Stay Focused

Some might say it is because we do not have the necessary will power to accomplish what we set out to do. Some say it is because we are too busy or too overwhelmed to take action on our resolution. My guess is it could be any of those things, but it is more likely that you have just started down a path without your compass and you have started to lose your way.

Rather than rattling off a list of things, you "should" do for whatever reason, sit down and think about what it is you want to achieve and set a solid intention for accomplishing your goal. I also suggest that you focus on only one or two intentions at a time. No matter what it is that you would like to achieve, setting an intention can and will set you on a course for success.

Tips for Staying Focused

Some might say it is because we do not have the necessary will power to accomplish what we set out to do. Some say it is because we are too busy or too overwhelmed to take action on our resolution. My guess is it could be any of those things, but it is more likely that you have just started down a path without your compass and you have started to lose your way.

Rather than rattling off a list of things, you "should" do for whatever reason, sit down and think about what it is you want to achieve and set a solid intention for accomplishing your goal. I also suggest that you focus on only one or two intentions at a time. No matter what it is that you would like to achieve, setting an intention can and will set you on a course for success.

Here are 5 top tips to finally achieving your goal:

- **Tip 1**

Get clear. In setting an intention, you're making it clear to yourself and other people exactly what you plan to do. Define the definition of what accomplishing your goal would be. For instance, you know you've reached your goal of improving your management skills when you consistently feel more satisfied with your ability to deal with tough situations and motivate your staff. You may even get that promotion you have been after!

- **Tip 2**

Realize that an intention comes in several sizes and every large goal is filled with intentions big and small. With follow-through, each intention will ultimately lead to success. For instance, if you resolve to improve your management skills, your first intention may be to speak with your company to find out what skills and traits you may want to focus on.

- **Tip 3**

do not let confusion overwhelm your intention. You may have lots of passion for your resolution, but passion without a plan is wasted energy and will eventually fizzle out. Setting an intention to take a step towards your goal each day will keep you on the right path and help to clear away confusion.

- **Tip 4**

Use your resources. Ask for what you want and need from other people. When you clearly state your intention and your request of other people, you have the opportunity to gain a partner and a cheering section. For instance, if you look up to somebody's management style, ask him or her for tips and possibly even support. Chances are they will be flattered and very willing to share advice.

- **Tip 5**

Be accountable. Choose your resolutions carefully by deciding what interests you. You might ask somebody you trust to help keep you accountable. Nevertheless, nothing can take the place of honoring your intentions to yourself. You will be amazed at how your self-esteem and sense of accomplishment will increase when you achieve your goals.

Chapter 19: Working Less Perform More

If you are feeling overworked, overwhelmed or just plain over it, the following time-management tips can help you maximize your productivity so you can accomplish more.

- **Work Less Accomplish More**

If you are feeling overworked, overwhelmed or just plain over it, the following time-management tips can help you maximize your productivity so you can accomplish more.

- **Separate Work from Home**

Between responding to personal emails, instant messaging and fielding cell phone calls from your kids, it can get very hard to stay focused on the tasks. Therefore, when you're in the office try to concentrate on your work as much as possible. Then when you're at home, you can deal with your issues there without distraction. You'll wind up having better quality time in both places. Separating your work duties from home-related ones will allow you to keep your mind on work when you're there and, in turn, procrastinate less, feel less overwhelmed and accomplish more.

- **Establish boundaries and stick with them**

While it is always great to try to make everybody happy all the time, it is just not possible in a workplace ruled by the irrefutable laws of time and space. Learn when to say no. There are times when it's right to go beyond the call of duty on the job. For Instance, when it is a real emergency, then I do not mind staying late or going out on a limb.

However, that's different from just letting people dump their last-minutes work on your desk so they can make it home early. While you need to do your work, you also need to take care of yourself and know your job's boundaries.

- **Get Organized**

Time spent hunting for files or lost phone numbers could be used for making progress on your to-do list. Good organizational structures are essential in any time-management plan. Spend a few moments at the end of every day answering voicemails, and emails. It always helps to be organized and not let messages pile up. It will always save you time. Sticky notes posted on your keyboard can help you remember the most important task that needs to be done throughout the day.

Everyone has a system for being organized. Try these tips. They may just add a couple of minutes to your day along with your routine you already practice.

- **Make Time for Yourself**

Any well-constructed to-do list has to include some time for relaxing and centering yourself, or you might wind up too stressed out to do anybody any good.

Hard work = Resistance and is the opposite of flow.

First off, no one likes to "have to" do anything. When you say I "have to" pay my bills, so I "have to" work hard to get the money. Already you can sense the despair and powerlessness in the very idea of "having to" do something you naturally resist. When you "have to" do anything, you're in a state of resistance. You are fighting the natural flow by pointing yourself upstream, resisting. After a few decades of this resistance, you can see how some individuals eventually burn out, lose their perspective, lose joy, create sickness and ultimately lose life itself. No, having to work hard to get what we want

isn't the answer. Grinding away at something is the root cause of all disappointments in life. Turn this idea around.

Discover and pursue your path of least resistance.

What do you love? I mean love! What do you like to "play at"? What are you effortlessly good at? What activity excites you to an extent that when you're engaged in it, you lose your awareness of time? Think about these questions deeply. Within this idea of "play" is the seed of joyous, easy, relaxed, natural, lazy creation. There's no working when you're engaged in an activity you feel you were born to do. If you have an idea that you love to play with, do you force yourself to play with it? Of course not. You are naturally drawn to it. You're lovingly engaged in it. Things are easy. There's no work involved and the results of your creation seem almost heaven-sent.

Your whole life must reflect what you're naturally drawn to do. It is essential to accomplishing your heart's desire. Do not trade one more second of your precious life energy working hard at achieving your goals. Discover your greatest gifts that have been with you since the day you were born and use them to create value in a simple and relaxed way! Everything you need to create your success is already within you.

Any useful idea that has elevated the life experience of individuals has come about because people would like to avoid having to do hard work. All our innovations throughout history have been created to make life easier and better. Hard work is counter-productive to the direction of growth and life-expanding. Hard work shuts off the flow of creative, inspired energy. Hard work isn't in alignment with the laws of creation. You're made of the same stuff and this natural law applies to you wittingly or unwittingly. You'll never become healthy, wealthy and wise:

- keeping your nose to the grindstone

- Pushing the ball uphill

- Working your fingers to the bone

- Going to a salt mine

- Spending the day with a slave driver

There's an easier, lazy, do-nothing way to create the life you have always desired. You must engage yourself in what you love, play and have fun with. Play with everything. If it is not fun and feels like hard work, you're decreasing your potential for creating massive success in your life. Align your focus and attention to only that which you love.

Then find partners who love doing the activities you resist doing. When you put it all together, you will take a quantum leap in your power to create what you desire.

Chapter 20: Self-Confidence

- **What is Self-Confidence?**

Philosophy teaches us to bear with equanimity the bad luck of other people.

Merriam-Webster defines equanimity as evenness of mind under stress - a habit of mind that's rarely disturbed under great strain; a controlling of emotional or mental agitation through will and habit; a steadiness when facing the strain.

Equanimity is a practice, most often discussed in Buddhist and Sufi traditions. Equanimity is the base for wisdom and freedom and compassion and love. Few individuals are capable of expressing with equanimity opinions that differ from the prejudices of their social environment. Most individuals are even incapable of forming such opinions.

- **What does Self-Confidence look like?**

Equanimity is the capacity to stay neutral, to observe from a distance, and be at peace without getting caught up in what we observe. It is the capacity to see the big picture with understanding and without reacting, for instance, to another's words, ideology, perspective, position, premise, or philosophy. Essentially, we take nothing personally; refuse to be caught up in the drama our own or other peoples.

Equanimity allows us to "stand in the midst," of conflict or crisis in a way where we are balanced, grounded and centered. Equanimity has the qualities of inner peace, well being, vitality, strength, and steadfastness. Equanimity allows us to remain upright in the face of the strong winds of conflict and crisis, such as blame, failure, pain, or disrepute - the winds that set us up for suffering when they begin to blow. Equanimity protects us from being "blown over" and helps us stay on an "even keel."

- **How do we develop Self-Confidence?**

Numerous mind/body qualities support the development of equanimity. One is integrity. Do-ing and be-ing in integrity support our feeling confident when we speak and act. Being in integrity fosters an equanimity that results in "blamelessness," feeling comfortable in any setting or with any group without the need to find fault or blame. Another quality that supports equanimity is faith (not necessarily a religious or theological faith) - a faith-based on wisdom, conviction or confidence. This type of faith allows us to meet challenges, crisis or conflict head-on with confidence, with equanimity. A third quality is that of a well-developed mind a mind that reflects stability, balance, and strength. We develop such a mind through a conscious and consistent practice of focus, concentration, attention, and mindfulness. A well-developed, calm mind keeps us from being blown about by winds of conflict and crisis.

A quality that supports equanimity is seeing reality for what it is, for instance, that change and impermanence are an unpleasant fact. We become detached and less clingy to our attachments. This means letting go of negative judgments about our experience and replacing them with an attitude of loving-kindness or acceptance and a compassionate matter-of-factness. The more we become detached, the deeper we experience equanimity. The final quality letting goes of our need to be reactive so we can witness, watch and observe without needing to get caught up in the fray, the winds - maintaining a consistent relaxed state within our body as sensations move through.

Equanimity, thus, has two aspects: the power of reflection and an inner balance, both of which support one to be mindful, awake, aware and conscious. The greater the degree we are mindful, the greater our capacity for equanimity. The greater our equanimity, the greater our ability to remain steady and balanced as we navigate through the rough waters and gusty winds of change, challenge, and conflict.

- **What happens when we are out of balance lacking Self-Confidence?**

In our everyday physical world, when we lose our balance, we fall. In our emotional world, we stuff our feelings and emotions, deny them or contract around them. Or we identify with a particular thought, feeling or emotion, hold on to it rather than allow it to flow through us or pass like a cloud in the sky. The middle ground is equanimity - the state of non-interference.

Equanimity allows for a deeper, more fulfilling experience.

As we develop our capacity for equanimity, we can begin to notice when we drop into a "state of equanimity." Being aware of our experience, we can explore the state and this practice will lead to more frequent and deeper states of equanimity. What we find with such practice is that people, events, and circumstances that once caused us to be reactive no longer have any "charge" and we are more and more able to let go and feel less "bothered." We suffer less.

Chapter 21 Use of Statements And Conclusion Finals

If you use affirmations, make it a point to use them frequently and do not stop using them even if your situation is getting better. The more you use affirmations, the better the situation. There are affirmations for every type of situation; priorities, procrastination, focus, and serenity are just a few. Here are some affirmations that may help you throughout your day.

- **Affirmations**

If you use affirmations, make it a point to use them frequently and do not stop using them even if your situation is getting better. The more you use affirmations, the better the situation. There are affirmations for every type of situation; priorities, procrastination, focus, and serenity are just a few. Here are some affirmations that may help you throughout your day.

- **Prioritizing**

I am entitled to live a calm life, full of joy and order.

I set realistic goals, remembering that my priority is myself.

I schedule tasks at the right pace for me.

I proactively decide what tasks I should do first and which are more important.

I take something off my schedule before I add one on.

I make time for anything new that I bring into my life.

I find no need to hoard my time on one specific thing.

I can delegate the tasks that I cannot - or should not - be doing.

I make time for play and rest and do not allow myself to work non-stop.

I accept my progress because it is at my own pace.

I know that my patience, tolerance, and efforts help me learn and grow to be a stronger version of myself.

I am gentle with my efforts, knowing that my new way of living requires much practice and patience.

- **Procrastination**

I take the action and leave the outcome up to God.

There is no try, only do.

My desire to get on with things is stronger than my desire to procrastinate.

Procrastination is my enemy.

I take action towards my goals daily.

I enjoy the success of finishing a task.

I choose to start this task with a small, imperfect step. I will feel terrific and have plenty of time to play!

I look forward to getting things done.

Among everything I do, I love working the most.

- **Focus**

I am alert and attentive at all times.

I am always focused on what I am doing.

I am attentive and observant throughout my day.

I am aware and present at all times.

I am calm and focused in all that I do.

I am completely absorbed in the present moment.

I am completely focused on what I am doing.

I am easily absorbed by all that I do.

I am fascinated and involved in every task I perform.

I am focused and relaxed in all that I do.

I am fully focused and present in all interactions with others.

- **Serenity**

I have Peace of mind at all times.

No matter what I am faced with I will remain calm.

I have inner peace.

My mind is always in a calm loving place.

- **Wrapping Up**

Generally, when we make progress in life it is because life presses on us to move forward. Seldom is it because of our own conscious choice and initiative? Over the long course of our lives, such forced advancement occurs in unpredictable ways, sometimes through happy, but often through unhappy experiences. Surely, this is not the most effective way to progress in life.

Yet that is the way the overwhelming majority of us progress in life. We call such inefficient, unpredictable, life-meandering progress. However, with this book and these tips, we can consciously change our behavior without the added stress.

We can have balance and success in life. Besides, we can get our priorities in line and keep them there.

What is stopping you from grabbing those extra minutes and hours in your day?

The only answer is, you! So get out of your way and start accomplishing more by doing less!

© Copyright 2019 by C. Baker - All Rights Reserved

www.ingramcontent.com/pod-product-compliance
Lightning Source LLC
Chambersburg PA
CBHW060850220526
45466CB00003B/1310